ENGINEERING ENERGY STORAGE

ENGINEERING ENERGY STORAGE

ODNE STOKKE BURHEIM

ACADEMIC PRESS

An imprint of Elsevier

Academic Press is an imprint of Elsevier
125 London Wall, London EC2Y 5AS, United Kingdom
525 B Street, Suite 1800, San Diego, CA 92101-4495, United States
50 Hampshire Street, 5th Floor, Cambridge, MA 02139, United States
The Boulevard, Langford Lane, Kidlington, Oxford OX5 1GB, United Kingdom

Notices

Knowledge and best practice in this field are constantly changing. As new research and experience
broaden our understanding, changes in research methods, professional practices, or medical
treatment may become necessary.

Practitioners and researchers must always rely on their own experience and knowledge in evaluating
and using any information, methods, compounds, or experiments described herein. In using such
information or methods they should be mindful of their own safety and the safety of others,
including parties for whom they have a professional responsibility.

To the fullest extent of the law, neither the Publisher nor the authors, contributors, or editors,
assume any liability for any injury and/or damage to persons or property as a matter of products
liability, negligence or otherwise, or from any use or operation of any methods, products,
instructions, or ideas contained in the material herein.

Library of Congress Cataloging-in-Publication Data
A catalog record for this book is available from the Library of Congress

British Library Cataloguing-in-Publication Data
A catalogue record for this book is available from the British Library

ISBN: 978-0-12-814100-7

For information on all Academic Press publications
visit our website at https://www.elsevier.com/books-and-journals

Working together
to grow libraries in
developing countries

www.elsevier.com • www.bookaid.org

Publisher: Joe Hayton
Acquisition Editor: Lisa Reading
Editorial Project Manager: Charlotte Rowley
Production Project Manager: Paul Prasad Chandramohan
Designer: Mark Rogers

Typeset by VTeX

CONTENTS

PREFACE

The front cover of this book illustrates the three branches of energy production, distribution, and consumption. These areas are overlapping by the field of energy storage. By the introduction of renewable energy technologies these three areas will become closer and more interconnected. This is so because most renewables are electric, intermittent, and often out of phase with the demand in energy consumption. It is energy storage that can balance this. Thus energy storage is a growing field for engineering and research. Understanding the elements of engineering energy storage is needed now and ever more in the future.

Energy is the backbone of our society. Today, we have a fossil-based economy. More than eighty percent of the energy consumption today is coal, oil, and gas. This is what drives our economy. It is fundamental in providing our society with the services needed, whether it is transportation, food production, health care, entertainment, or any other. However, the threat from climate changes from the emissions from our economy backbone, coal oil, and gas, means that we need to make changes. Fortunately, the usage of fossil energy is so inefficient that we only need to replace a third of this energy if a renewable energy technology is chosen. That is, it takes at least 3 MWh annually of mixed coal, oil, and gas to make a MWh of electricity each year, though a MWh of renewable energy annually is a MWh annually exactly—and once installed it is without marginal costs.

Renewable, carbon neutral, and sustainable are keywords essential for the energy economy and energy technology of the future.

Inevitably, the energy supply must come from renewable sources, and this change has started.

Regardless of chosen technology, the technology itself must be carbon neutral. This means that if carbon is needed for the technology process or manufacturing, then it must be taken form the air, not the ground. For example, carbon is needed for producing metals like silicon, aluminum, and steel or for components in Li-ion batteries, fuel cells, or supercapacitors. Solutions of extraction this from plants and trees exists and relies more, in my opinion, on political will than technological barriers.

Finally, the installed solutions must be sustainable, both with respect to the environment footprint and for the growth of the economy of our society. Currently, we can find examples all over the world where the fear of climatic unsustainable use of fossil energy has led to much needed incentives for renewable energy by wind and solar. Installing vast amounts of such intermittent energy technologies needs to be accompanied by buffering solutions, not only fiscal incentives to adjust consumption patterns. We need vast amounts of energy storage solutions on the one hand and energy storage solutions of many different types for many different purposes on the other.

This book is intended to help engineering students before moving to a master thesis level, already trained engineers, and interested scientists, with understanding the engineering concepts behind the most relevant energy storage technologies.

Several people deserve a big thank you for supporting me, believing in me, and taking me in on a professional side: Signe, Preben, Jon, Bert, Frode, and, for the longest period, Håvard. You have all been great role models for me in your own very different ways. Ms. K. Taksdal is acknowledged for help with proof reading and with a critical eye from a students perspective.

On a personal side, a big thank you to all members of the family I grew up with, especially to my parents.

A very special thank you goes to my wife for her patience with me in writing this book, of which most of the writing took place during the first seven months after receiving our first child.

CHAPTER 1

Energy Storage

Energy in the right form, at the right place, and at the right time is the key essence here. Energy drives our economy and is founded on engineering these three aspects. Thus, when installing more electric renewable energy, energy storage technologies that conserve this energy into an adequate form with adequate distribution properties for use whenever needed is what it is all about.

There are many different energy storage technologies that in their own different ways facilitate the trinity of properties: formate, distribution, and timing. As a part of the transition to a renewable energy economy, a great variety of energy storage technologies must be taken into use. This book explains the theory behind the different technologies and why this gives the different technologies different advantages and markets.

As a renewable energy-based economy is inevitable, engineering energy storage is inevitably needed too. This is now.

1.1 A BRIEF HISTORY OF ENERGY

All animals look for energy that they eat. This energy is stored either in vegetables nearby or in animals that live nearby the predators. Almost all of this energy comes from sunlight, that is, in combination with water and chlorophyll, converted to hydrogen via the photosynthesis. In turn, when combined with CO_2, hydrocarbons of various kinds are produced to store this energy. In this sense, carbon is a hydrogen carrier, and hydrogen is an energy carrier.

An important part of the evolution of not only *homo sapiens* but of our entire society is our ability to stack up energy for later. Storing food is a first example of energy storage, but the humans are not alone in this context. Several animals store food for later. The other very important development in the history of humans was the ability to make and control fire. This is the first example of controlled energy conversion that we know about in nature. It was important for human evolution since it allowed us to digest meat much more efficiently and grow larger brains and next start with houses and agriculture. One key element here is storing energy in the form of food and dry wood. These two forms of energy are chemical energy. Whereas

Engineering Energy Storage.
DOI: 10.1016/B978-0-12-814100-7.00001-8

the first was used for nutrition, the latter was used for heat generation, both whenever needed. The advantage of chemical energy is that it is extremely easy to store. One can keep it in a bunker or a tank. Other forms of energy are less convenient to store, and the quality is sometimes degraded.

Society developed further as villages and cities emerged. Mechanical energy in the form of mills deploying water to rotate a grain mill is a first example. During the first industrial revolution, coal started to be used in steam engines, converting heat into work. At the end of this era, electrical energy started to be developed too. We can already now address four important forms of energy: chemical, mechanical, electrical, and thermal. Chemical energy can be stored on tanks. Mechanical energy can be stored at elevation, rotation or pressure. Thermal energy can be stored by insulation. Electrical energy cannot be stored without being converted into one of the other forms. Over the last centuries, our society has been organized in such a way that mechanical and chemical energy is supplied when needed. This pattern, in particular for the electrical energy, is now about to change.

Over the last few decades, with the introduction of wind and photovoltaic (PV) energy conversion, our society has changed from supplying energy on demand to supplying energy when energy is present, that is, when the sun shines and the wind blows, electric energy is generated and supplied regardless of demand. Also, this energy production does not have much of a cost margin, meaning that once the machines are there, they generate electric energy almost for free. Traditional energy production requires coal, oil, natural gas, uranium, and, to some extent, manpower. Wind and PV hardly do. When energy is supplied beyond the demand, it needs to be stored for later. It is the impact of the intermittent, no-cost-margin, large-scale, and world-wide energy production that will drive the need for energy storage.

Another very interesting change presently occurring is the localization of the energy production and its connection to the Internet. Energy consumers can install a PV on their roof (or in their garden) and thus become energy suppliers, at least for certain hours of the day. This energy production changes over the year in addition to the hours of the day. The energy unit (as the house now becomes) sometimes produces electric energy and sometimes consumes electric energy. This needs to be handled one way or another. Locally, the electricity is distributed on the grid, for instance, to the neighbors or nearby supermarket. It is the transition

Table 1.1 A brief historical over view of development in the modern industrialized society

Ind. Rev.	Field and characteristics		
When	Energy:	Transport:	Telecommunication:
1st: 1750–1950	1750–: Digging coal from the ground to run steam engines and later oil and gas for electrical power plants. The energy passes through big hubs and is subsequently redistributed.	1800–: Trains and ships carry people and freight over large distances. For transport, one relies on going via large stations.	1890–: Telephone allowed for direct communication over large distances. This happened via central hubs that connected you via regional and local centrals.
2nd: 1950–	2010–: Everyone can produce electricity and sell it to their energy enterprise, but the energy is delivered locally. Peer-to-peer energy distribution.	1950–: The common man can buy a car and drive from door to door or further. Localized and customized transportation.	1990–: Internet and smart phones. People communicate directly over the Internet. Peer-to-peer.

from having electricity generated at large-power plants into a localized energy market that represents the big change. The change from having a commodity distributed from the big banks, centrals, and hubs to a more localized market is often acknowledged as a second industrial revolution.[1] We have in the past seen this for the transport and telecommunication sectors as common people got cars and tools to communicate peer-to-peer online via chat rooms. This is illustrated in Table 1.1. The impact of the second industrial revolution in the energy sector is not yet quite understood. It will, however, require more storage of electric energy for days, weeks, and months. Moreover, the energy infrastructure already developed will not become redundant. From the so-called second industrial revolution in the transportation sector and the telecommunication sector we can see that this second industrial revolution only contributes to the infrastructure that was already there. In less developed parts of the world, however, this new infrastructure of local electric energy becomes the only one.

---

[1] Historians consider several industrial revolutions, while here it is simplified to two.

1.2 RENEWABLE ENERGY AND ENERGY STORAGE

Renewable energy and power sources are getting cheaper and competitive with existing ones (coal, oil, gas, and nuclear). Once installed, these power sources have hardly no operational costs. This has led to a growing share of the renewable energy sources. Because the renewables currently growing in market share (wind and PV) have intermittent nature and represent the cheapest alternative when available, the established power sources are seeing significant changes in the need for them. That is, they must deliver more power for some hours during a day and hardly none for other hours of the day. These changes are demanding because the existing power plants were built for stationary power production and with a perspective of 30–40 years. The introduction of renewable energy is not only lowering CO_2 emissions; it has in two to five years changed the electric energy market that we have known for decades. The characteristics of these changes continue to grow in strength. The most sensible ways to ensure that the electric energy market remains sustainable is by ensuring hourly power rates, appropriate political solutions, and including a lot of different energy storage solutions.

The final impact and politics for implementation of renewable energy in the form of wind and PV are difficult to predict. However, there are a few indications already revealing themselves. One is the introduction of electricity meters that allows for price timed with the hours of the day, as an economic political mean to curb the consumption peak, particularly in evening hours. However, this can be seen as late coming in regions where PV are installed and continue to be installed at the consumers location. This creates a deficit in the demand for electric power from the grid when the sun shines. In many regions, the consumers have a legal right to become suppliers to the grid too. With consumers becoming suppliers that force energy with no marginal (operational) cost into the grid or just not demanding any, new market scenarios emerge. Some trends remain as expected; however, the impacts are different. In Fig. 1.1, we can see some of these impacts from continuously adding PV at the power grid end point. Some changes are as expected and others are not.

Traditionally, two trends are generally accepted as a nature of the electric power market, a market that will be strongly affected by the implementation of renewable energy sources. One is that the energy consumption has always been increasing. The other is that there are peaks in the power consumption in the morning and in the evening. When we evaluate these statement in the light of Fig. 1.1, we can see that the power demand from the grid,

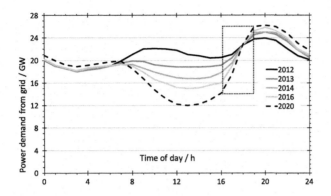

Figure 1.1 Electric grid power demand during a day in January in California [1].

in the absence of solar power from PVs, is increasing every year. When looking at the power supplied from the main grid in 2012, we can also see the morning and evening peaks. This has been for decades. The reason for these peaks is that residential power consumption goes up as people are making food and using different utilities in the morning before going to work and in the evening after returning from work. Whereas the evening peak is growing annually, the morning peak is not only removed, it has turned into a dropping hill over only two years. Note that the valley on the graph continues to sink and is estimated to continue lowering further as more and more residential and other buildings are installing PVs.

It is not the power consumption that is lowered in the morning and daytime; it is the demand for electricity from the main grid that is lowering in this time frame. In fact, the ratio between the minimum and maximum power from the central grid changes by more than a factor of two. This means that since 2020, the daily fluctuations in the power need from central power units (coal, oil, gas, nuclear, and wind) doubles in the time frame of three hours. Whereas the wind power is delivered at nearly no operating costs and therefore goes into the base line supply (along with nuclear), the remaining power sources are going to tune up their capacity several times.

It can be argued that every renewable energy unit put into the market is a good thing within the aspect of greenhouse gas emissions. However, the market still relies on these traditional power machines to support the ever growing evening peek. Moreover, having large fossil-based power stations idling for hours during day time leaves useless CO_2 emissions and is also expensive for the owners. In fact, much of the basis for investing into these types of power plants vanishes or at least changes completely. This change became apparent over two years and continues to hit stronger during the following five years. The immediate solution to keep the electric energy market sustainable, keeping investors for large-scale power plants willing to continue investments, is to increase the price of electricity several times during night time and to compensate the idling during day time. From an environmental point of view, sustainability can be increased by introducing more localized and highly efficient energy storage systems that comply with the PV power production and large-scale energy storage systems that can support the central grid properties. Incentives like hourly priced electricity are still required. In this way, distributed energy storage can lower the average and maximum power demand, and the large-scale power stations can be fewer and produce power more efficiently and in a reliable and predictable market.

Distributed and efficient energy storage will primarily consist of lithium-ion batteries and phase change heat storage tanks for electric and thermal energy. For the centralized large-scale energy storage devices, several types will be needed: pumped hydro, compressed air, hydrogen, flow batteries, fly wheels, and more.

This book is intended to describe the theory needed to engineer the demand for power and energy, to understand the system size capability, and to understand the main cause and propagation of energy efficiency in cycle, single step, and chains of energy storage devices.

1.3 ENERGY AND POWER

In the field of energy storage, there are two important terms, energy and power. Power refers to how much energy in the form of work is supplied in a given time frame. The car analogy terms to power and energy are horse powers and kilometers, respectively. Power is the rate of energy. This is shown in Eq. (1.1), where W denotes energy in the form of work, P is power, and t is time:

$$P = \frac{dW}{dt} \Leftrightarrow W = \int_{t_1}^{t_2} P dt. \tag{1.1}$$

There are many examples of energy systems that offer a lot of power but little energy. This means that they only supply the given power for a short time. There are also many systems that do the opposite, they contain a lot of energy, but the rate of energy delivered is for some reason limited.

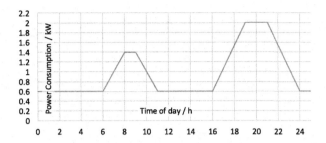

Figure 1.2 A typical power consumption at home on a week day.

When designing an energy storage system, it is the balance between the energy capacity and the power capacity that must be accounted for. One must make sure that the system can deliver enough energy and enough power. This is the starting point. In many instances, the capacity of the power and energy comes at different prices. Moreover, high power is often only needed for shorter periods like in the morning and afternoon at home or while driving uphill with a car. Therefore, one must often choose a combination of two systems and make a hybrid one. When doing this, one opts for enough of the power intensive component to deliver enough power and then choose a system with lower power output but with adequate capacity to support it.

Example 1.1: Power and Energy Need at Home.

The power consumption of a home on a weekday is shown in Fig. 1.2. The home has a lot of PV on the roof, and the idea of the owner is to be self-supplied with electricity. This is to be done with to energy storage systems, hydrogen and batteries. The concern or problem formulation here is how can we have a combined energy system that can store enough energy for the time that the sun does not shine. This system must be able to deliver both energy and power. For this, we are going to use hydrogen for energy capacity and batteries for power capacity. Moreover, due to season variations in sunlight, the system needs to be designed for 8 hours of sunlight (09.00–17.00). The surplus of energy is stored as hydrogen from electrolysis, which is returned via fuel cells. The battery is also charged during day time. The fuel cell system must deliver enough power for the average energy consumption. During 6 hours throughout the night, the battery must be charged from the fuel cell system so that it can support the morning power peak. Since the battery is charged from PV (solar cells) during daytime, the recharge during night is the bottleneck for the proposed hybrid energy system. In designing this storage system, a stepwise recipe is given:

a) What is the electric energy consumption each day (in kWh)?

b) What is the required power need for the fuel cell, considering 100% energy efficiency, in order to match the average power need?

c) What is the required power for the water electrolyzer?

d) What power capacity must the battery package have to match the gap during peak hours?

e) How much energy does the battery need to obtain during night hours?

f) What is this in power and is the power gap from the fuel cell during night sufficiently large?

Solution:

a) By using Eq. (1.1) we can calculate or integrate the energy for every hour and summarize. This is done in the following table and corresponds to 23.2 kWh:

Time	0–6	6–8	9–11	11–16	16–19	19–21	21–24	Total
Energy (kWh)	3.6	2.0	2.8	3.0	3.9	4.0	3.9	23.2

b) The required power need for the fuel cell is the average energy per time:

$$P = \frac{\Delta W}{\Delta t} = \frac{23.2 \text{ [kWh]}}{24 \text{ [h]}} = 0.967 \text{ kW}.$$

c) The electrolyzer must supply the required fuel cell energy during 8 hours. The fuel cell operates at 0.967 kW for 16 h. Thus the power of the electrolyzer must be no less than

$$P_{e.lyzer} = \frac{E_{F.C.}}{t_{e.lyzer}} = \frac{0.967 \text{ [kW]}16 \text{ [h]}}{8 \text{ [h]}} = 1.93 \text{ [kW]}.$$

d) The largest power required is 2 kW. This is 1.033 kW more than the fuel cell delivers. Hence the battery package must deliver 1.033 kW.

e) The battery has to deliver 0.2 kWh (07.00–08.00) and 0.4 kWh (08.00–09.00), which is 0.6 kWh in total.

f) From about 23.00 to 07.00 the battery is being charged. The fuel cell supplies close to 0.4 kW for 6 hours and on average 0.2 kW for 2 hours, and thus the energy for charging is close to 2.8 kWh. This is more than enough energy for the problem here.

In Example 1.1, the house is considered as an off-grid solution where the energy is supplied from unlimited access to PV electricity between 9 and 17. One can argue whether this is a likely example, however it is possible. Another exercise one can do is to consider potential future energy deals. By many it has been proposed that electricity will become a subscription base on quality rather than quantity, or an upper power (kW) boundary limit with otherwise free access to energy (kWh). This would be somewhat similar to how Internet was delivered to homes in

Table 1.2 Specific energy and power for transportation

	Specific power /kW kg^{-1}	Specific energy /kWh kg^{-1}
Fuel cells (2017)	0.25–1.5	0.5–1.5
Li-ion batteries	0.15–0.25	0.1–0.5
Super capacitors	2–4	0.002–0.005
Steam engine on a train	0.2–0.3	0.5–2
Boeing 777 jet engine	5–10	3–10

the 1990s and 2000s. In the 1990s, one would pay for every bit downloaded, whereas in the 2000s, one would pay only for the bandwidth, that is, bit/s. As renewable energy sources are growing and a need for controlled costumer demand rises, a scenario where electricity subscription is a monthly fixed rate as long as one never uses more than, for instance, 1 kW in one's home is a reasonable business model. This would fit with the average energy need of the home in Example 1.1 and cover the growing valley in Fig. 1.1, in addition to shaving of the evening demand peak. In this instance, it can become interesting to add a battery package like in the example. The battery numbers will be different, although the means to solve the dimensionalization of the problem is almost similar.

1.3.1 Energy and Power for Transportation

For most transport applications, weight is a very important factor in choosing an energy system. This is in particular valid for personal cars. Having a system that offers a combination of energy, power, and low weight is important. Obviously, price, efficiency, and volume are also important factors, but mass appears to be the merit by which energy and power is evaluated. Therefore it is much more relevant to use the terms specific power and specific energy, that is, power and energy per mass. The common way to present specific power and energy is called a Ragone chart. In such a chart, one plots specific energy on one axis and specific power on the other. Which one is on the primary and secondary axis is of less importance, but it is important that both axes are logarithmic. That is because the Ragone chart evaluates the properties by order of magnitude. Some examples of specific energy and power are given in Table 1.2. These are also presented as a Ragone chart in Fig. 1.3.

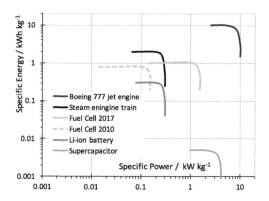

Figure 1.3 Ragone chart-based on the values in Table 1.2.

Comparing the values in Table 1.2 to those in Fig. 1.3, it appears that the maximum potential of the technology is not displayed. For instance, the Li-ion battery is suggested to be capable of 0.3 kW/kg and 0.3 kWh/kg at the most, but this value is not indicated in the Ragone chart. Instead, the maximum point of energy density is plotted together with the suggested minimum specific power and vice versa. A line defined by these two points and a third point, that has values consisting of above intermediate value of both specific power and energy is used as a curving point. The energy devices can never have the maximum power density and maximum energy density at the same time. This fact is related to the fact that high power utilization leads to lower energy efficiency and thus lower actual specific energy in that situation. This goes for almost all energy converters. Thus the lines define a maximum set of available specifc power and energy for each of the technologies.

A remark to this Ragone chart, in comparison to older ones, is that the fuel cell offer more specific power and energy than a Li-ion battery. In Ragone charts that are older, this is not the case, and therefore fuel cell properties as of 2010 and 2017 are shown.

1.3.2 Efficiency and Propagation of Efficiency Losses

Efficiency is by definition the ratio of a lesser outcome from a larger input. It does not have to be the ratio between work in and out; it can also include heat. It can be for power (rate of energy) or it can be for energy. In most cases, however, this book deals with work as the output that we consider as the output of interest, and usually work is also the input property of interest. For work only, the energy efficiency ε is defined as

$$\varepsilon = \frac{W_{out}}{W_{in}}. \tag{1.2}$$

Efficiency also propagates. For several processes in series, where energy is stored and retrieved and possibly also converted, one can determine the total energy efficiency by multiplying the efficiency for each of the terms. This is as in Eq. (1.3). Here we also show an example of the energy storage train consisting of AC/DC conversion–electrolysis–fuel cell–DC/AC conversion.

$$\varepsilon_{chain} = \left(\frac{W_{DC}}{W_{AC}}\right)_{AC/DC}\left(\frac{\Delta G_{H_2}}{W_{DC}}\right)_{e.lyzer}\left(\frac{W_{DC}}{G_{H_2}}\right)_{F.C.}\left(\frac{W_{DC}}{W_{AC}}\right)_{DC/AC},$$

$$\varepsilon_{chain} = \varepsilon_{AC/DC} \cdot \varepsilon_{e.lyzer} \cdot \varepsilon_{F.C.} \cdot \varepsilon_{DC/AC},$$

$$\varepsilon_n = \prod_{i=1}^{n} \varepsilon_i. \tag{1.3}$$

Multiplying all the energy efficiencies of a train of conversion units with each other, we obtain the energy efficiency for the entire system. This means that it is better to convert utilization of an energy source in two steps losing, for example, 10% in each step than to do one conversion losing, for example, 20%. However, the extra capital cost of a two-step solution often leads to a less efficient single-step solution.

Example 1.2: Accounting for Energy Efficiency.
In Example 1.1, we looked at the requirements for storing energy in the form of hydrogen. We learnt that the required power of the electrolyzer was 1,93 kW and the fuel cell power requirement was 0.967 kW. Most electric devices require a AC voltage of 230 V (in Europe). Under the consideration that an electrolyzer has an energy efficiency of 82%, a fuel cell has an energy efficiency of 85%, a DC/AC converter for home usage has an energy efficiency of 97%, and that energy lost in relation to storage leads to a storage efficiency of 96%; what is the overall energy efficiency and what is the required power for the electrolyzer in Example 1.1?

There are two questions to be answered. The query about the efficiency is resolved by using Eq. (1.2):

$$\varepsilon_{syst.} = \varepsilon_{e.lyzer}\varepsilon_{storage}\varepsilon_{F.C.}\varepsilon_{DC/AC} = 0.82 \cdot 0.96 \cdot 0.85 \cdot 0.97 = 65\%.$$

Summarizing the losses (0.18 + 0.04 + 0.15 + 0.03), we would end up with an efficiency of 60%. This would be wrong. But the exercise tells us, as is already mentioned in the text, that energy efficiency propagates underproportionally, meaning that an energy loss in a chain of energy losses has a reduced impact compared to its single value.

The other question about the need for the power to of the electrolyzer is answered by using Eq. (1.2):

$$\varepsilon_{syst.} = \frac{W_{AC}}{W_{e.lyzer}} = \frac{P_{AC}.t_{F.C.}}{P_{e.lyzer}t_{e.lyzer}}$$

$\updownarrow$

$$P_{e.lyzer} = P_{AC}\frac{t_{F.C.}}{t_{e.lyzer}}\frac{1}{\varepsilon_{syst.}} = 0.967\,[\text{kW}]\frac{8\,[\text{h}]}{16\,[\text{h}]}\frac{1}{0.65} = 2.98\,[\text{kW}].$$

This means that due to the energy losses, we need about 50% extra PVs to cover the energy needs of the of-grid house in Example 1.1.

PROBLEMS

Problem 1.1. Energy and Power in Cars.
A car of mass 1500 kg brakes from 50 km/h in 3.5 seconds when meeting a red light. Based on the intermediate values in Table 1.2, determine
a) the mass needed for a supercapacitor and a Li–ion battery.
b) the distance the technology can take you at 50 km/h (requires 10 kW).

Problem 1.2. Energy Efficiency and Remote Storage.
A common figure used for energy loss in the electric grid is 8%. Suppose that water is pumped up at night time with an efficiency of 95% at night time and released at an efficiency of 85% in the morning.
a) What is the energy efficiency of sending the electric power to the energy storage facility at night and to have it returned in the morning?
b) What is the largest loss in energy and what is its relative contribution in the cycle?

SOLUTIONS

Solution to Problem 1.1. Energy and Power in Cars.
a) First, we must determine the kinetic energy. At 50 km/h, this is

$$E_{kin} = \frac{1}{2}mv^2 = 0.5 \cdot 1500\,[\text{kg}]\left(\frac{50\,[\text{km/h}]}{3.6\,[\text{ks/h}]}\right)^2 = 145\,[\text{kJ}] = 0.040\,[\text{kWh}].$$

The power thus becomes

$$P = \frac{\Delta E}{\Delta t} = \frac{145 \ [\text{kWs}]}{3.5 \ [\text{s}]} = 41 \ [\text{kW}].$$

The mass needed on an energy perspective is the given as the energy divided by the specific energy:

$$m_i = \frac{E_i}{e_i}.$$

With respect to the power, the mass is given by the power need divided by the power density:

$$m_i = \frac{P_i}{p_i}.$$

Intermediate values and corresponding weight needs are:

Technology	Specific energy /[kWh/kg]	Mass need /[kg]	Specific power /[kW/kg]	Mass need /[kg]
Supercapacitor	0.0035	11	3	14
Li-ion battery	0.35	0.11	0.2	205

To take up the energy from braking, the supercapacitor and Li-ion battery must have masses of 11 and 0.1 kg, respectively. To take up the power, they must have masses of 14 and 205 kg, accordingly. Therefore, this braking requires 14 kg of supercapacitor or 205 kg of battery. Clearly, the supercapacitor is the lighter choice.

b) When the car travels 50 km/h, a power need of 10 kW is suggested. We need to relate the energy in the storage device to the energy and the driving distance:

$$Energy \ need = Energy \ on \ board,$$
$$P_{need} \Delta t = e_i m_i,$$
$$P_{need} \frac{l}{v} = e_i m_i \Leftrightarrow l = \frac{v}{P_{need}} e_i m_i.$$

Thus we can calculate the driving distance for the given technologies:

$$l_{supercap.} = \frac{50 \ [\text{km/h}]}{10 \ [\text{kW}]} 0.0035 \ [\text{kWh/kg}] 14 \ [\text{kg}] = 245 \ [\text{m}],$$

$$l_{sLi-ion \ B} = \frac{50 \ [\text{km/h}]}{10 \ [\text{kW}]} 0.35 \ [\text{kWh/kg}] 205 \ [\text{kg}] = 359 \ [\text{km}].$$

Thus we see that a battery package has the capacity to brake a car from 50 km/h to still stand in 3.5 second and, at the same time, travel of 360 km in an urban area, which explains why Li-ion batteries are such a success in small short-range (200 km) vehicles.

Solution to Problem 1.2. Energy Efficiency and Remote Storage.

a) We must first make an overview of the energy efficiency of each step in the series:

	Grid night	Pumping	Discharge	Grid morning
ε	0.92	0.95	0.85	0.92

The overall energy efficiency is given by Eq. (1.3) as

$$\varepsilon = 0.92 \cdot 0.95 \cdot 0.85 \cdot 0.92 = 68\%.$$

b) The largest contribution to the loss in energy efficiency is the discharge. Comparing all the relative losses (not efficiencies), we get its relative contribution from the discharge $\frac{15}{8+5+15+8} = 22\%$.

CHAPTER 2

General Thermodynamics

Thermodynamics is a very useful tool for any engineer when analyzing processes. Although one might find it theoretical, it is actually a very applied field. It is applied so frequently and extensively and in so many fields, that most fields have their own ways to treat thermodynamics. Physicists typically evaluate an atom and extrapolate this into a system. Mechanical engineers usually consider a continuum and account for mass. Chemists and most material scientists typically account for thermodynamics in terms of moles ($6.02 \cdot 10^{23}$ particles). Since energy storage is closely related to stored chemical energy, most of the thermodynamics in this book will account for moles, the chemical approach. In other instances, we use the mechanical engineering approach. The readers are likely to have different backgrounds, and therefore, in this chapter, we give a brief summary of some introductionary thermodynamics.

2.1 THE FIRST LAW AND INTERNAL ENERGY U

For any closed system, we can define the first law of thermodynamics by saying that the heat added Q minus the work extracted W must equal any change in kinetic energy E_{kin} and gravimetric potential energy E_{pot}:

$$\Delta E_{pot} + \Delta E_{kin} = Q - W. \qquad (2.1)$$

Any other change is then defined as a change in internal energy U. We define a change in internal energy of a closed system as any other change after we have added heat, extracted work, and changed kinetic and potential energy:

$$\Delta U = Q - W - \Delta E_{pot} - \Delta E_{kin}. \qquad (2.2)$$

The internal energy depends on the pressure and temperature. The potential energy depends on the elevation and mass. The kinetic energy depends on the mass and velocity. When an energy property of a substance depends on the state that it is in, it is called a state function. Potential energy, kinetic energy, and internal energy are state functions, whereas work and

Engineering Energy Storage.
DOI: 10.1016/B978-0-12-814100-7.00002-X
Copyright © 2017 Elsevier Inc. All rights reserved.

heat are process functions, making the heat and the work are very different from the other three. Whereas these three others are state functions and depend on the state that they are in (pressure, temperature, mass, height, and velocity), the heat and work are process variables since their size or amount depend on the way that something is done. This is exemplified in Example 2.1.

Example 2.1: Heat or Work.

For a certain pressure increase in a piston, is heat or work required?

When quickly pushing a piston into a bicycle pump, the pressure increases. The next thing you will notice is that heat is released and the piston pushes less back. An alternative way to increase the pressure is to heat up the piston without adding any work at all. Because the pressure and temperature change is the same in the two processes and the pump is at rest and at the same elevation before and after, the change in internal energy is the same in both cases. In other words, it does not matter for a state function how a change is done, but it matters for the process functions heat and work.

In most cases that we evaluate in this book, the system evaluated is at rest and does not change potential energy, and we can define the energy balance as

$$\Delta U = Q - W. \tag{2.3}$$

Equation (2.3) is fit to describe thermodynamic cycles for a given closed system like a piston expanding and contracting while exchanging heat and work with the surroundings. However, in many instances, we are interested in systems where mass enters and leaves continuously when heat is constantly added or work is constantly extracted. This situation demands a different form of Eq. (2.3). When mass enters and leaves the substances entering the system, then some energy, which is neither heat nor work, is given by the state of the substance entering or leaving kinetic, potential, or internal energy:

$$\dot{Q} - \dot{W} = \dot{m}_{out}\left(u_{out} + gz_{out} + \frac{v_{out}^2}{2}\right) - \dot{m}_{in}\left(u_{in} + gz_{in} + \frac{v_{in}^2}{2}\right). \tag{2.4}$$

The dot over some of the terms indicates that there is a rate, and the lower case u is the specific (per mass) internal energy.

2.2 SECOND LAW AND ENTROPY

Entropy S is a thermodynamic property that is somewhat hidden on a daily basis and therefore is difficult to grasp. When we evaluate work and heat, these are process properties that we experience very directly on a day-to-day basis. Entropy, however, is an underlying property relative to the present temperature. Regardless of its abstract properties and delusive appearances, entropy can be evaluated and used as a very important tool for engineers. Here we describe entropy in three different forms: transported, Q/T, changed, ΔS, and produced, σ. Just like with energies, these forms are linked and can be summarized, but a very different one, the production of entropy σ is always greater than or equal to zero since the energy is always conserved. So one of the definitions of the second law of thermodynamics is that the entropy *production* cannot be negative:

$$\sigma \geq 0. \tag{2.5}$$

Entropy can be very concrete when comparing likelihood of different configurations. For instance, ordered accumulation is less likely to appear than random distribution, and thus the former represents lower entropy compared to the latter. This is a very common form of entropy description, and the change in order is a good metaphor for a change in entropy, ΔS.

When evaluating entropy as a thermodynamic property, it quickly appears to be much more abstract. For thermodynamical processes, entropy is definitively related to heat. Evaluating *reversible* heat, meaning a process that is independent of the process route, we define the relation

$$dQ_{rev} = TdS. \tag{2.6}$$

Accordingly, a reversibly added amount of heat is defined as $Q_{rev} = T\Delta S$. This is the heat transferred to the system.

Example 2.2: Heat Transport Is Entropy Production.

Consider two bodies of the same material at different temperatures, 300 and 400 [K]. Heat (1200 [J]) is transported spontaneously from the hotter body to the colder one. Consider the temperature of the two reservoirs to be unchanged. The heat is conserved, however, not the entropy.

The entropy of the hotter body decreases: $\Delta S = \frac{-1200 \text{ [J]}}{400 \text{ [K]}} = -3$ [J/K].

The entropy of the colder body increases: $\Delta S = \frac{1200 \text{ [J]}}{300 \text{ [K]}} = 4$ [J/K].

We can see that the entropy changes are not conserved, so the transport of heat produces entropy, in this instance, 1 [J/K]. Because heat spontaneously and inevitably always goes from a hot to a cold place, entropy is spontaneously produced, meaning that it happens by itself.

However, most of us have experienced that heat is generated in a process without being added, and the most common example from daily life is friction. Rubbing ones hands against each other, this heat is felt. We give some work by engaging our muscles, and this work is then lost as heat. This generation of heat is nonreversible. So friction is often referred to as lost work. This heat relative to the temperature it is generated at is called the entropy production σ. Because it is a result of heat generated or produced within the system, it is called the produced entropy.

We have now introduced three types of entropy: (i) entropy changed in a system, ΔS, (ii) entropy transferred to a system, Q/T, and (iii) entropy produced in a system, σ. These entropies are related by the entropy balance, Eqs. (2.7) and (2.8), and express the entropy change as a function of heat added and produced within the system:

$$dS = \int_{H}^{C} \frac{dQ}{T} - \sigma, \tag{2.7}$$

$$\sigma = \sum_{i=1}^{n} \frac{Q_i}{T_i} - \Delta S \geq 0. \tag{2.8}$$

In Sections 2.2.1 and 2.2.2, we present two consequences, which are *very* important for engineers.

2.2.1 Reversible Adiabatic Must Be Isentropic

An important fact about the second law is that the entropy production is always greater than or equal to zero. A more practical one is that the entropy change depends only on temperature and pressure or on temperature and specific volume. Entropy production and entropy transferred depends solely on the process route. The most relevant practicality of the entropy balance and Eq. (2.8) is that when heat is not added (or removed) and no friction occurs, the entropy does not change. Therefore, processes where heat is not added, adiabatic ones, when reversible (friction free), are isentropic. **Reversible adiabatic processes are isentropic**. We return to this later in this chapter.

2.2.2 The Carnot Efficiency Limitation

Converting heat to work is a very interesting area and still a growing research area. In particular, the lower hanging fruit of low-grade waste heat utilization has received increase interest. However, there is a limitation to how much work one can extract. This limitation is called the Carnot efficiency and is dictated by the second law of thermodynamics and derived form the entropy balance.

Considering a wall that does nothing except conducting heat, the heat in and heat out are the same, that is, $Q_H = -Q_C = Q_{in}$. Since there is no friction or change of entropy in this process, Eq. (2.7) can be rewritten as

$$\sigma = \frac{Q_C}{T_C} - \frac{Q_H}{T_H}. \tag{2.9}$$

Considering a wall that actually produces some work, like, for example, a Peltier module, the heat out is different from the ingoing, defined by the extracted work: $Q_H = Q_{in}$ and $Q_C = Q_{out} = Q_{in} - W_{out}$. Equation (2.7) now becomes

$$\sigma = \frac{Q_{in} - W_{out}}{T_C} - \frac{Q_{in}}{T_H}. \tag{2.10}$$

We now can write an expression for the relation between work out and heat into a process:

$$\frac{W_{out}}{Q_{in}} = \left(1 - \frac{T_C}{T_H}\right) - \frac{T_C}{Q_{in}}\sigma. \tag{2.11}$$

The best case scenario according to the laws of thermodynamics is that the entropy production σ is zero. This gives the maximum work obtainable from a heat source, but more importantly, we term this efficiency the Carnot efficiency ε_{Carnot}

$$\frac{W_{max\ out}}{Q_{in}} = \left(1 - \frac{T_C}{T_H}\right) = \varepsilon_{Carnot}. \tag{2.12}$$

Example 2.3: The Entropy Production of Added Work.

Increased work extraction leads to a lowered entropy production. A lowered work output ultimately stems from an increased entropy production.

a) Rewrite Eq. (2.10) to express the entropy production for a wall where electric work is added (e.g., heat in a radiator).

b) What is the upper and lower limits for the entropy production when adding work like this?

Solution:

a) *The heat out is now increased, so that from Eq. (2.9) we can write:*

$$Q_{out} = Q_{in} + W_{in},$$

$$\sigma = \frac{Q_{in} + W_{out}}{T_C} - \frac{Q_{in}}{T_H}.$$

b) *The lower value is defined by the entropy production of heat going through the wall: $\frac{Q_{in}}{T_C} - \frac{Q_{in}}{T_H}$. There is no upper limit in how much entropy the system can produce by adding work into the wall.*

2.3 PRESSURE AND VOLUME

In Section 2.2, the reversible (and nonreversible) heat are introduced.

If, on the other hand, we consider a force F acting on a thermally insulated piston with a cross-sectional area A so that the piston moves a distance dx, then we can define the reversible work dW_{rev}:

$$dW_{rev} = Fdx = \frac{F}{A}Adx = pdV. \tag{2.13}$$

We have now used the first law of thermodynamics in a differential form to define changes of internal energy for systems at rest, by development of Eq. (2.3), as a function of two states (T and p) and two state functions for different state functions (S and V):

$$dU = TdS - pdV. \tag{2.14}$$

Example 2.4: Heat or Work?

We can increase the pressure of a cylindrical piston by one mean at the time. Think, for instance, of a bicycle pump.

a) Which two means?

b) Evaluate the change in entropy in the two instances.

c) What does this simple experiment tell us about reversible entropy changes?

Solution:

a) *We can compress the piston so quickly that heat is incapable of exchanging with the surroundings, **or** we can add heat at constant volume.*

b) *Let us first consider the **heating at constant volume**. The density of particles is not changed, but energy is added, and the temperature increases. Since the par-*

ticles, due to the temperature increase, have greater velocity and bounce around faster, they are also less ordered. Thus the entropy increases.

Let us next consider the **adiabatic compression**. During the swift compression, molecules are compressed, and the temperature increases. The compression calls for a decrease in entropy and the heat for an increase. By definition, Eq. (2.6), for an insulated or adiabatic process, the change in entropy is zero, simply because Q is zero. The internal energy of the gas in the piston still changes, and solely due to the added work.

c) For reversible processes, heat is required to have entropy changes. If there is only work added, only the internal energy changes in response.

We have established that heat is required for entropy to change in reversible processes. We have seen that this can happen at constant volume, also called isochoric.

$$dS = \frac{dU}{T}\frac{dT}{dT} - pdV \underset{isochoric}{\Rightarrow} \left(\frac{dU}{dT}\right)_V \frac{dT}{T}. \tag{2.15}$$

To develop Eq. (2.15) further, we need to introduce two new thermodynamic relations. The heat capacity for constant volume is defined by the heat accumulated as internal energy relative to temperature increase:

$$C_v = \left(\frac{dU}{dT}\right)_v = \frac{TdS}{dT}. \tag{2.16}$$

Adding work alone changes the internal energy, temperature, and pressure together with the volume. An interesting question subsequently rises: *How does temperature and volume relate for adiabatic processes?*

Developing Eq. (2.15) to answer this question requires introducing the ideal gas law. The ideal gas law is the relation between pressure, volume, and temperature for a given amount of matter and the gas law constant:

$$pV = n\bar{R}T = mRT \Leftrightarrow p = \frac{mRT}{V}. \tag{2.17}$$

The molar ideal gas law constant $\bar{R}$ is defined as 8.314 J/mole K independently of substance. It can be defined as a specific ideal gas constant R, the molar divided by the molar weight of the subject substance.

We can now define the entropy change for a process where we first change the temperature and then later the volume:

$$dS = C_v d\ln T - mR \, d\ln V. \tag{2.18}$$

This is interesting for evaluating states of adiabatic or isentropic ($dS = 0$) processes, now the compression by volumetric figures and the initial temperatures. Classical examples are piston processes, where air is suddenly heated by a combustion process, i.e. heat added so fast that the volume can not expand, and then expanded isentropically at constant pressure afterwards. By integration we obtain:

$$\ln \frac{T_2}{T_1} = \frac{R}{C_v/m} \ln \frac{V_2}{V_1} \Leftrightarrow T_2 = T_1 \left(\frac{V_2}{V_1}\right)^{\frac{R}{c_V}} \Leftrightarrow \frac{T_2}{T_1} = \left(\frac{v_2}{v_1}\right)^{\frac{R}{c_V}}. \qquad (2.19)$$

where c_V is the specific heat capacity at constant volume.

As discussed in Example 2.4, a gas can experience pressure increase by adding heat, work, or both. When compressing a gas very quickly, for instance, a piston in an internal combustion engine, heat does not have time to leave. Under reversible conditions, the entropy ballance tells us that there is no change in entropy either and thus the process is adiabatic. In this instance, Eq. (2.19) can be rewritten as

$$\frac{v_2}{v_1} = \frac{v_{r,2}}{v_{r,1}}, \qquad (2.20)$$

where $v_{r,i}$ is not really a volume; it is rather a fictive or relative volume used to manoeuvre in adiabatic compression tables. An example is given.

Example 2.5: Adiabatic Volume Compression.

Starting with 100 L of air in a piston at 25°C, what is the temperature when the gas is compressed to 50 L, 10 L, and 1 L?

Use the gas compression table in Appendix B.

Answer:

Since this is a closed volume, Eq. (2.20) can as well be $\frac{V_2}{V_1} = \frac{v_{r,2}}{v_{r,1}}$, and we get $v_{r,2} = \frac{V_2}{V_1} v_{r1}$. To solve the problem, by insertion we get the following table:

V_2/L	100	50	10	1
v_{r2}	100	50	10	1
$T/°C$	25	115	445	1287
$u/kJ\,kg^{-1}$	213	278	526	1261

First, we determine the relative "volume" $v_{r,2}$. Then we look to table in Appendix B and find these volumes. From these rows and by interpolation we get the other requested data. From this we can also determine the specific work required: 65, 461, 800 kJ kg^{-1}.

2.4 ENTHALPY AND CONTROL VOLUMES

When evaluating control volumes, like heat exchanger, valves, turbines and compressors, a gas can expand inside the fixed controlled volume. Because pressure changes rather than the volume, the work is no longer related to pdV but rather to Vdp. This calls for different measures. The work is thus defined as $W = W_{C.V.} - Vdp$. Inserting this into Eq. (2.4) at stationary state, we obtain:

$$\dot{Q} - \dot{W}_{C.V.} = \dot{m}_{out}\left(u + vp + gz + \frac{v^2}{2}\right)_{out} - \dot{m}_{in}\left(u + vp + gz + \frac{v^2}{2}\right)_{in},$$
(2.21)

where v is the specific volume. The vdp term is a property change for the gas inside a control volume that is not directly work in the same way as we are used to think for pistons. After accounting for changes in kinetic energy, potential energy, heat added, work from the control volume and internal energy, something more happens. This something more is Vdp. This is a property of the state that needs further investigation. By summarizing the internal energy and the pressure volume term, we define the enthalpy

$$H = U + pV.$$
(2.22)

By partial differentiation and insertion of Eq. (2.14) we obtain:

$$dH = dU + pdV + Vdp = TdS + Vdp.$$
(2.23)

We can use this for three different things: i) we can define the energy balance as a function of enthalpy, ii) we can define the heat capacity for constant volume processes, and iii) we can define the temperature–pressure relation for adiabatic isentropic processes in open control volumes.

The energy balance is the most useful since one often finds enthalpy tabulated or in databases:

$$\dot{Q} - \dot{W}_{CV} = \dot{m}_{out}\left(h + gz + \frac{v^2}{2}\right)_{out} - \dot{m}_{in}\left(h + gz + \frac{v^2}{2}\right)_{in}.$$
(2.24)

Since the reversible heat is defined as the temperature multiplied by the entropy change, Eq. (2.23) for a temperature response at constant pressure gives:

$$\left(\frac{dQ_{rev}}{dT}\right)_p = \left(\frac{dH}{dT} - V\frac{dp}{dT}\right)_p = \left(\frac{dH}{dT}\right)_p = C_p. \tag{2.25}$$

In turn, we can now rewrite Eq. (2.23) by replacing volume using the ideal gas law:

$$dS = \frac{dH}{T}\frac{dT}{dT} + Vdp = \frac{dH}{dT}\frac{dT}{T} + mRT\frac{dp}{p}. \tag{2.26}$$

For isentropic (reversible adiabatic) processes, we then obtain:

$$C_p \ln\frac{T_2}{T_1} = -mR\ln\frac{p_2}{p_1} \Leftrightarrow T_2 = T_1\left(\frac{p_1}{p_2}\right)^{\frac{R}{c_p}}, \tag{2.27}$$

where c_p is the specific heat capacity at constant pressure. Also note how the minus symbol is removed by inverting the ratio of the two pressures. For air, constituting of 21% oxygen and 79% nitrogen, the specific heat capacity at constant pressure c_p and constant volume c_v are 1.006 kJ K^{-1} kg and 0.717 kJ K^{-1} K^{-1} at room temperature and 1 atmosphere [2].

Likewise to in Section 2.3, we can transform Eq. (2.27) into a pressure relation where we determine a pressure that is not actually a pressure, but more like a relative pressure p_r:

$$\frac{p_2}{p_1} = \frac{p_{r,2}}{p_{r,1}}. \tag{2.28}$$

Example 2.6: Adiabatic Pressure Expansion.

A kg of air at 25°C and 1 atm passes through a compressor adiabatically. What is the exhaust temperature and enthalpy when compressing to 10, 50, and 1100 atm? (Use the nearest values and gas tables of Appendix B.)

Solution:

Equation (2.28) can be rewritten as $p_{r,2} = p_{r,1}\frac{p_2}{p_1}$. Looking at table in Appendix B, we get the following table:

	1	10	50	1100
p_2	1	10	50	1100
$p_{r,2}$	1	10	50	1100
$T/°C$	25	570	880	1850
$h/kJ\,kg^{-1}$	298	576	911	2065

From this we can also require the necessary work: 278, 613, and 1767 kJ.

2.5 GIBBS FREE ENERGY AND CHEMICAL POTENTIAL

Gibbs free energy represents the potential work from a chemical reaction and is defined as

$$G = H - TS. \tag{2.29}$$

By differentiation and insertion of Eq. (2.23) we obtain

$$dG = dH - TdS - SdT = -SdT + Vdp. \tag{2.30}$$

Equation (2.30) is valid only for closed systems, which quite often is less relevant. Often, we are interested in accounting for the effect of chemical reactions or the energy of mixing different components. This is when we introduce chemical potential

$$dG = -SdT + Vdp + \sum_i \mu_i dn_i. \tag{2.31}$$

Experimentalists, and particularly chemists, prefer to measure potential work at constant temperature and pressure. Chemists are interested in potential energy or work from chemical reactions. Hence, the term chemical potential μ at a given temperature, pressure, and a state of component mixture is defined by

$$\mu_i = \left(\frac{\partial G}{\partial n_i} \right)_{T,p,n_{j \neq i}}. \tag{2.32}$$

This means that the chemical potential for a component μ_i is defined as equal to the molar Gibbs free energy $\bar{g}_i$ at some given temperature, pressure, and mixture. For an ideal gas, we define as a reference pressure, a standard pressure p^o, where we also have a standard chemical potential μ_i^o. We now have a standard molar Gibbs free energy $\bar{g}_i^o$ at a standard pressure that we can define at any temperature T. Using the relation (from Eq. (2.30)) $d\bar{g} = d\bar{h} - Td\bar{s} - \bar{s}dT$ at any given temperature, we get

$$d\mu = d\bar{g} = d\bar{h} - Td\bar{s} = d\bar{h} - T\bar{c}_p \ln T + \bar{R}T \, d \ln p, \tag{2.33}$$

$$\Delta\mu = \Delta\bar{g} = \Delta\bar{h} - T\Delta\bar{s}^o + \bar{R}T \ln \frac{p}{p^o}, \tag{2.34}$$

$$\Delta\bar{g} = \Delta g^o + \bar{R}T \ln \frac{p}{p^o}. \tag{2.35}$$

Moreover, from its definition (Eq. (2.32)) the chemical potential is a molar property. It can therefore be added in molar mixtures on a component basis and by using Dalton's model and ideal gas mixtures (the sum of all partial pressures equals the total pressure and the partial pressure: $p_{tot} = \sum_i p_i$; for ideal gases, the partial pressure equals the molar fraction x_i times the total pressure):

$$\mu_{mix} = \sum_i x_i \bar{g}_i = \sum_i x_x \left(\bar{g}_i^o + \bar{R}T \ln \frac{p_i}{p^o} \right) \tag{2.36}$$

$$= \sum_i x_x \left(\bar{g}_i^o + \bar{R}T \ln \frac{x_i p_{tot}}{p^o} \right). \tag{2.37}$$

The energy of mixing is then the difference between the chemical potential and standard chemical potential:

$$\mu_i - \mu_i^o = \bar{g}_i - \bar{g}_i^o = \bar{R}T \ln \left[x \frac{p_{tot}}{p^o} \right] \tag{2.38}$$

or, in a differential form between mixed conditions with constant total pressure p_{tot},

$$d\mu_i = d\bar{g}_i = d \left(x_i \bar{R}T \ln [x_i] \right). \tag{2.39}$$

Because this potential energy is derived from the entropy term, it is often referred to as the entropy of mixing. This entropy turns into produced entropy σ if not taken advantage of. (River water mixing into the ocean is an example of such, where around 2 TW worth of potential, (mechanical or electrical, energy is continuously dissipated into the ocean [3]. This energy can be extracted in many ways; e.g. [4].)

In the continuation of thermodynamics and in nonideal mixtures, the mole fraction term x_i is replaced by activity. The activity of a component a_i is a relative concentration or fraction term that can be tailored to describe components in all kinds of systems. We return to this in greater detail in Chapter 6. The bottom line of this chapter is therefore that the chemical potential of any component i in any system at any temperature can be described by

$$\mu_i = \mu_i^o + \bar{R}T \ln a_i. \tag{2.40}$$

PROBLEMS

Problem 2.1. Dumb and Not so Dumb Walls.

1200 J of heat is transported through a wall. On one side, the temperature is 400 K, and on the other, it is 300 K.

a) How much work is potentially lost in this process?

b) Replacing the wall with a Peltier element that utilizes 45% of the entropy production for electricity production, how much electric work comes out of the process?

Problem 2.2. Work for Adiabatic Compression.

Determine the new temperature and specific work needed to at room temperature:

a) compress 100 L of air in a closed cylinder to 10 L;

b) pressurize air from 1 bar to 10 bar in a compressor.

SOLUTIONS

Solution to Problem 2.1. Dumb and Not so Dumb Walls.

a) If the process was entirely reversible, the entropy change would be the same in both of the reservoirs since that goes with no entropy production. Inevitably, 1200 J leaves the hot reservoir, and this thus dictates the lowest possible entropy change at the lower temperature (zero entropy production and reversible conditions means that the entropy change at higher temperature dictates the lowest possible entropy change at any lower temperature):

$$\Delta S = -\frac{Q_{rev}}{T} = -\frac{1200 \ [\text{J}]}{400 \ [\text{K}]} = -3 \ [\text{J/K}].$$

This means that the minimum heat that must be delivered to the colder side is

$$Q_{rev} = -T\Delta S = 300 \ [\text{K}] \, 3 \ [\text{J/K}] = 900 \ \text{J}.$$

From energy conservation ($W = Q_{in} + Q_{out}$), potentially, 300 J of work is lost in the process of transporting heat through this wall.

b) The entropy production in the example above is 1 J/K. Since 45% of this is now made use of, only 3.55 J/K comes out of the wall. This means that the energy balance along with the entropy balance gives us

$$W = Q_{in} + Q_{out} = Q_{in} - T(\Delta S + \sigma)$$
$$= 1200 - 300 \ [\text{K}] \, (3 + 0.55) \ [\text{J/K}] = 135 \ [\text{J}].$$

Solution to Problem 2.2. Work for Adiabatic Compression.
For pistons, we look at how volumes are lowered and use the internal energy changes. For compressors and turbines, we look at how pressure and enthalpy changes. In Appendix B, the adiabatic air compressibility table is given.

a) We find the values from the table:

V/L	T/K	$u/kJ\,kg^{-1}$	
100	298	213	$\Delta u/kJ\,kg^{-1}$
10	725	532	$= 319$

b) We find the values from the table:

V/L	T/K	$h/kJ\,kg^{-1}$	
1	298	298	$\Delta h/kJ\,kg^{-1}$
10	575	581	$= 283$

CHAPTER 3

Mechanical Energy Storage

3.1 INTRODUCTION

Mechanical energy storage, at its simplest, is something that has been done for a very long time. The first example is about 300 years after the wheel was invented. In doing pottery, one needed to have some stabilized momentum for making axis-symmetric products like bowls, cups, vases, plates, etc., and the solution was to have the pottery wheel centered on top of a rod, which in turn was placed in the center of a rotating stone cylinder at the bottom. All pieces could then be rotated around a vertical axis, and this was done by kicking the lower stone disk. In doing so, energy could be added in portions and buffered so that a steady momentum was available at the pottery wheel on top, where a piece of clay was quickly shaped into the desired object.

This old example of a very short-term energy storage is relevant today as well. The momentum of a rotating shaft in an internal combustion engine is another example. The piston needs to have work supplied to compress air upon ignition. After the ignition, much more mechanical energy is delivered back to the rotating shaft. When driving a manually transmitted car, one ramps up the kinetic energy of the shaft in the engine before carefully stepping of the clutch. At least, this is part of explanation. Other examples of stored kinetic energy also relate to rotation, since this is the only way to keep motion where one wants it.

3.2 MECHANICAL ENERGY STORAGE

In classic mechanics, there are two forms of energy, kinetic and potential. The typical example to start with is a pendulum that experiences no friction so that the energy in the system is conserved. By conservation, energy is given by the equation

$$E_{Mech.} = E_{pot.} + E_{kin.} = mgh + \frac{1}{2}mv^2, \tag{3.1}$$

where m, g, h, and v are the mass, gravitational constant, height available for work, and the velocity of the pendulum. The example of the pendulum

Engineering Energy Storage.
DOI: 10.1016/B978-0-12-814100-7.00003-1
Copyright © 2017 Elsevier Inc. All rights reserved.

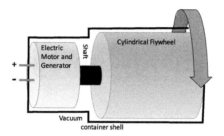

Figure 3.1 General sketch of an energy storage flywheel unit, consisting of an electric motor and generator, connected to the flywheel via a shaft inside a low-pressure vessel.

can be treated in much greater detail, but here we will look at energy storage for kinetic and potential energy in separate ways. The by far two most common solutions for storing energy solely by mechanical energy are electric flywheels and hydroelectric systems. These devices convert energy between electric and mechanical energy.

3.2.1 Flywheels

A flywheel is a cylindrical or disk-shaped object that stores energy by rotating around its center axis. When storing small amounts of energy, the wheel typically has a thickness smaller than the radius and is then disk-shaped. When storing large amounts of energy, the thickness is larger than the diameter, and it is then cylindrically shaped. In principal, a flywheel energy storage system needs three components: an electric motor, the cylinder/disk, and a shaft that connects these two. It is the cylinder/disk that is the flywheel. When storing energy, the motor acts as a load on the electrical grid. When delivering energy, the electrical motor becomes a generator and acts as a power source. To store energy for longer, the flywheel is encapsulated inside a container that is evacuated. This is to lower friction from the shear forces in the air surrounding the flywheel. A sketch of a flywheel system in a container is shown in Fig. 3.1. Depending on the materials of the flywheel, rotational speed of up to 100 000 rpm can be applied. This can cause enormous friction – even for air.

3.2.1.1 The Energy

In studying the amount of energy in a flywheel, the starting point is imaging a very thin cylindrical shell that rotates around its center axis. This shell is illustrated in Fig. 3.2. With the kinetic energy of a moving object defined

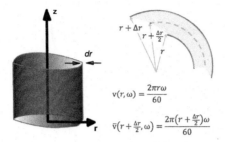

Figure 3.2 The cylindrical shell geometry used for deriving the energy of a cylindrical shell.

by the equation

$$E_{kin} = \frac{1}{2}mv^2 = \frac{1}{2}\rho V v^2,\tag{3.2}$$

we can define the kinetic energy of the rotating cylindrical layer as a function of the inner and outer radius r, height h, density ρ, and rotational velocity ω.

The velocity of any given point in the shell is given by the equation

$$v = 2\pi r \frac{\omega}{60},\tag{3.3}$$

where r is the radius, and ω is the rotational velocity in rounds per minute (rpm). Since the kinetic energy changes as the radius increases, one intuitive way to evaluate the energy is by its variation, that is, its radial derivative. One classical approach in this problem is to state that the shell has an inner radius r and an outer radius $r + \Delta r$. Based on this and Eqs. (3.2) and (3.3), we can define the kinetic energy of the outer and inner layers and then allow for the thickness of the layer to become infinitely small. This gives us the energy in a differential form so that we can later find the energy of the entire body by integrating. Finding the partial energy requires finding the limit as Δr approaches zero. First, we define the energy change. In this analysis, we use the simplification that all the kinetic energy is gathered along the average radius of the cylindrical shell and that the representative average velocity is $\bar{v} = 2\pi \left(r + \frac{\Delta r}{2}\right) \frac{\omega}{60}$, as indicated in Fig. 3.2

$$\Delta E_{kin} = \frac{1}{2}m\left(\frac{2\pi(r + \frac{\Delta r}{2})\omega}{60}\right)^2\tag{3.4}$$

$$= \frac{1}{2}\rho h(A_{shell}) \left(\frac{2\pi(r + \frac{\Delta r}{2})\omega}{60} \right)^2. \tag{3.5}$$

The area of the bottom of the cylindrical shell, A_{shell}, is defined by π times the squared difference between the outer and inner radiuses, $((r + \Delta r)^2 - r^2)$. With the radius sensitivity simply being Δr, we obtain the radial derivative of the rotational energy:

$$\frac{dE_{kin}}{dr} = \lim_{\Delta r \to 0} \frac{1}{2} \frac{\rho \pi h \left((r + \Delta r)^2 - r^2\right) \omega^2 \left(2\pi(r + \frac{\Delta r}{2})\right)^2}{60^2 \Delta r} \tag{3.6}$$

$$= \lim_{\Delta r \to 0} \frac{1}{2} \frac{\rho \pi^3 h \left((2r(\Delta r) + (\Delta r)^2)\right) \omega^2 4 \left(r^2 + 1/2r\Delta r + (\frac{\Delta r}{2})^2\right)}{60^2 \Delta r} \tag{3.7}$$

$$= \lim_{\Delta r \to 0} \frac{1}{2} \frac{\rho \pi^3 h (2r + \Delta r) \omega^2 4 \left(r^2 + 1/2r\Delta r + (\frac{\Delta r}{2})^2\right)}{60^2} \tag{3.8}$$

$$= \frac{4\rho \pi^3 h r^3 \omega^2}{60^2}. \tag{3.9}$$

Solving this first-order nonlinear differential equation by integration, we obtain an expression for a cylindrical shell with inner and outer radii:

$$E_{kin} = \int_{r_1}^{r_2} \frac{4\pi^3 \rho h \omega^2}{60^2} r^3 \, dr = \frac{\pi^3 \rho h \omega^2}{60^2} \left(r_2^4 - r_1^4\right). \tag{3.10}$$

Two common engineering problems are to find the rotational energy of a cylinder or of a cylindrical shell. These are specialized problems of Eq. (3.10), and therefore they also have specialized forms of the equation:

$$E_{kin} = \frac{\pi^3 \rho h \omega^2}{60^2} r_{cyl.}^4, \quad \text{(3.10)a} \qquad E_{kin} = \frac{\pi^3 \rho h \omega^2}{60^2} \left(r_o^4 - r_i^4\right). \quad \text{(3.10)b}$$

For the first problem, the rotating cylinder, we can use Eq. (3.10)a and that the radius of the cylinder is $r_{cyl.}$. This equation can be representative for fly wheels storing energy in combination with an electrical motor/generator like that illustrated in Fig. 3.1.

For the other problem, a rotating rim (cylindrical shell), we can use Eq. (3.10)b with outer and inner radii r_o and r_i. At first, this type of geometry might seem less relevant, but it is important for parts stabilizing machines. Examples can be transmitting wheels or additional mass on crank shafts for internal combustion engines with few pistons or at low rotational rates.

Neither of the two equations (3.10)a,b accounts for elastic energy at high rotational speed, allthough for some special cases this can be up to ten percent [5].

3.2.1.2 Other Aspects

There are many things to consider when it comes to flywheels.

Perhaps the most important one is the power. As seen from Eq. (3.10), the energy is related to radius, rotation, and density. On the other hand, the power is related to two components, the electrical motor and the shaft between the motor and the cylinder. It is important to notice that the energy and power capacities are determined by different component properties.

Moreover, the upper limit for how much energy that a material can store depends on the ratio between its density and tensile strength. If this ratio is too low, then the material disintegrates by the centrifugal forces imposed by the rotation. Most flywheels are limited to some 20–30 krpm, but some very special materials can take up to 80 or almost 100 krpm without disintegrating [5].

When rotating, centrifugal forces are imposed. This leads to stretching of the material. As long as this happens within a regime where the material is not exposed to forces outside the E-modulus of the material, that is, the region where the material is elastic (phenomenologically, like a rubber band), the stretching causes no trouble. We must of course take the stretching into account when designing the size of the low-pressure chamber so that the disk does not touch the walls under operation. The elastic stretching of a flywheel also represents additional stored energy. When a flywheel is rotating at its peak velocity, the elastic component can account for as much as ten percent of all the energy stored [5].

Because the rotation imposes these centrifugal forces on the flywheel itself, most high-rpm flywheels are not shaped like cylinders. They are in fact often shaped more dish or soup bowl like (imagine two plates facing each other), where they have a rather thick height near the center axis that gradually decreases as the radius increases. This has two classical engineering aspects: one is that the material closest to the center axis is less exposed to the centrifugal forces, and the other, more typical one, is that it takes material to carry material. We call this an overlinear mass demand. A classic example, of this overlinearity, is the turbine blades of a wind turbine. The amount of material is almost proportional to the square of the length of the blade. This is because the inner part of the blade must be much thicker than for shorter blades to carry the extra weight of more blades, and blades that are exposed to higher centrifugal forces (further out gives higher sentrifugal force at a given rotational speed ω.

Example 3.1: Energy in a Discos-Shaped Flywheel.

A flywheel is disc shaped somewhat like a discos. The height changes with the radius, so that the height is 10 cm at the center axis and 3 cm at the outer radius, which is 15 cm. The thickness decreases linearly with r. The flywheel rotates at 16 krpm, and the material has a density of 3 000 kg m^{-3}. Neglect energy contributions from elasticity.

a) Show that the stored energy can be expressed as

$$E_{kin} = \int_{0m}^{0.15m} \frac{4\rho_0 \pi^3 h\omega^2 (r^3 - 4.67r^4)}{60^2} dr$$

$$= \left[\frac{4\rho_0 \pi^3 h\omega^2 \left(\frac{1}{4}r^4 - \frac{4.67}{5}r^5 \right)}{60^2} \right]_0^{0.15m},$$

where ρ_0 is the density of the material of the flywheel.

b) What is the stored energy? Give the answer in kWh.
c) What power is this if released in 20 seconds?
d) What is the mass of the disk?
e) What specific power does this refer to if released in 20 seconds and the auxiliary components mass is 20 kg?
f) What factor change in power and specific power does this refer to if the rotation was 75 krpm instead of 16 krpm and the density half?

Solution:

a) From Eq. (3.8) we have:

$$\frac{dE_{kin}}{dr} = \lim_{\Delta r \to 0} \frac{1}{2} \frac{\rho \pi^3 h (2r + \Delta r)\omega^2 4 \left(r^2 + 1/2r\Delta r + (\frac{\Delta r}{2})^2 \right)}{60^2}. \quad (3.11)$$

The key to solving this problems lies in rearranging its formulation. Instead of assuming that the thickness of a material with constant density changes, we will assume that the density of rotating disk with a constant thickness changes. The density changes relative to the thickness so that it has hundred percent of the density at the center (0.1/0.1) and then a relative decrease of $\frac{0.1-0.03}{0.15\cdot0.1}$ times the radius. The density thus becomes

$$\rho = \rho_0 \left(\frac{0.1}{0.1} - \frac{0.1 - 0.03}{0.15 \cdot 0.1} r \right) = \rho_0 (1 - 4.67r), \quad (3.12)$$

where ρ_0 is the density of the material itself, 3 000 kg m^{-3}. By inserting Eq. (3.12) into Eq. (3.11) we obtain:

$$\frac{dE_{kin}}{dr} = \lim_{\Delta r \to 0} \frac{1}{2} \frac{\rho_0 (1.0 - 4.67r) \pi^3 h \, (2r + \Delta r) \, \omega^2 4 \left(r^2 + 1/2 r \Delta r + (\frac{\Delta r}{2})^2 \right)}{60^2}.$$

In turn, this becomes

$$\frac{dE_{kin}}{dr} = \frac{4\rho_0 (1.0 - 4.67r) \pi^3 h r^3 \omega^2}{60^2}, \quad 0\,\text{m} \le r \le 0.10\,\text{m}. \tag{3.13}$$

By integrating we obtain:

$$E_{kin} = \int_{0m}^{0.15m} \frac{4\rho_0 \pi^3 h \omega^2 (r^3 - 4.67 r^4)}{60^2} dr$$

$$= \left[\frac{4\rho_0 \pi^3 h \omega^2 \left(\frac{1}{4} r^4 - \frac{4.67}{5} r^5 \right)}{60^2} \right]_0^{0.15m}. \tag{3.14}$$

Alternatively, one can introduce an expression for the height change as a fucntion of the radius and keep the density constant. The answear will be the same.

b) *By inserting the given values we obtain:*

$$E_{kin} =$$
$$\frac{4 \cdot 3,000\,[\text{kg m}^{-3}] \pi^3 0.1\,[\text{m}] 16,000^2\,[\text{min}^{-2}] \, (0.25 \cdot 0.15^4\,[\text{m}^4] - 0.933 \cdot 0.15^5\,[\text{m}^4])}{60^2\,[\text{s}^2\,\text{min}^{-2}]}.$$

Note that the slope coefficient 0.2 has the unit m^{-1} and that when inserting the inner boundary condition (0 m), there is nothing to subtract for.

$$E_{kin} = \frac{9.53 \cdot 10^{12} (5.57 \cdot 10^{-5})}{3600}\,[\text{kg m}^2\,\text{s}^{-2}] = 147\,[\text{kJ}] = 41.0\,[\text{Wh}].$$

41 Wh is actually not too much energy. That is why this needs to be placed in context of the mass of the flywheel and the conversion time.

c) *The power is the energy released of a period of time:*

$$P = \frac{\Delta E_{kin}}{\Delta t} = \frac{41\,[\text{Wh}]}{0.00556\,[\text{h}]} = 7.38\,[\text{kW}].$$

Now, 7.38 kW is a significant amount of power. It is almost 10 hp, however, not that much for a car.

d) *The mass of the disk is found by integration:*

$$m = \int \rho dV = 2\pi \rho_0 h \int_0^r r - 4.67r^2 dr$$

$$= \left[2\pi h\rho_0 \left(\frac{r^2}{2} - \frac{4.67r^3}{3} \right) \right]^{0.15m} = 11.3 \text{ [kg]}.$$

e) *In addition to the wheel itself, there are the auxiliary components (electric motor and shaft) of said 20 kg. A mass of 31.3 kg gives a specific power of 0.235 kW kg^{-1}. The specific energy can also be calculated, 1.31 Wh kg^{-1}. These two numbers are really important for evaluating suitability in the transport sector. Looking at Fig. 1.3, we have designed a system that has the energy density slightly larger and power density equivalent to that of a typical super capacitor. If considering a 75 kW electric engine of mass 100 kg instead of 20 kg, the energy and power density would drop by almost a factor of five. Thus the auxiliary components are the most important factors for energy and power density in this example.*

f) *Finally to the what if this and that problem. This is a classical engineering problem of evaluating changes to a system. We are now to change the rotational velocity from 16 to 75 krpm and the density from 3,000 to 1,500 kg m^{-1}. Looking at Eq. (3.14), we can that as long as the energy is proportional to the square of the rotational velocity and directly proportional to the density,*

$$E_{kin}|_{r,h(r)} \propto \rho\omega^2.$$

Therefore, finding materials that can rotate faster is a good idea, even thought they have lower density. From the proportionality relation we get:

$$E_2 = E_1 \frac{\rho_{0,2}}{\rho_{0,1}} \left(\frac{\omega_2}{\omega_1} \right)^2 = E_1 \frac{1}{2} \left(\frac{75}{16} \right)^2 = 11 E_1.$$

The energy increases by a factor of eleven. So does the power as long as the discharge time is the same. The specific energy and power increase more because of the reduced weight, a factor of 11 compared to the initial. Therefore, as long as weight matters, reducing the density of the flywheel is irrelevant as long as it can rotate faster. Bearing in mind that even though we can improve the specific energy of the flywheel, the auxiliary components (electric motor and shaft) determine the available power and specific power of the energy storage system device.

3.2.2 Hydroelectric Energy Storage

As a mean to store vast amounts of energy for long times, hydroelectric power systems appear to be one of few technologies. Water is transferred from one height to another. When there is a surplus of electric energy on the grid, water can be elevated by pumping. This is done with an electric engine connected to a pump. When there is a shortage of electric energy, the water can be transferred down again through a turbine with a generator. This is done primarily by the use of a Francis turbine coupled to a generator. The Francis turbine is unique in the way that it can be reverted and operate under moderate to high pressure drops.

> **Example 3.2: Electric Motor or Generator.**
> On the next occasion that you come by a LEGO techno kit, pick up two of the engines from the kit and one lead. Couple these together by plugging the leads two ends on the electric LEGO socket on each of the engines. Turn the wheel pin on one of the engines (it is easier with a wheel on the pin) and look at the pin on the other engine. What happens?
> *The other pin moves at a lower rate. This is an example that electric engines can act as both engines requiring electric work and as a generator giving electric work.*

An electric motor can be reverted to act as an electric generator and vice versa. When it comes to pumps and turbines, two types of turbine types can be reverted, the Kaplan and the Francis ones. There are several other turbine and pumping technologies as well. The Pelton turbine and different displacement pumps are probably the most commonly known ones, however these are not revertible and have no relevance in the context of pumped hydro for energy storage. A Caplan turbine is perhaps best described as a propel like those on an outboard engine. This turbine is used in large-volume low-pressure streams like, for instance, in a river where a dam gives rise to some few meters of water height. A Francis turbine is quite complex in design, but when assembled, it looks a bit like a snails' shell. The turbine wheel consists of two parallel plates separated by thin walls in the radial direction that divides the space between the two plates into cake-shaped volumes. Around this cake-shaped double disk, the turbine wheel is the snail shell-shaped water distribution system. The turbine wheel typically rotates around the vertical axis, and water leaves through a hole in the center of the lower disk. The center of the upper disk is the attached to a shaft, which in turn is connected to the generator. A picture of such a Francis turbine wheel is shown in Fig. 3.3, right. In the picture, the wheel is upside down compared to how it is in operation. It is actually very common to be upside down old Francis wheel that is put on public display, most likely because the twists and turns of the blades are better displayed when upside down next to the ground. This way curious people

Figure 3.3 Pictures of a Pelton turbine wheel (left) and a Francis turbine wheel (right).

walking by can stick their arms through the flow compartments and have interesting pictures taken. When adding electric work to the generator, it becomes an engine, and the turbine becomes a pump. The cake defining walls between the two disks are usually curved for improved performance and the snail shell-shaped distribution manifold contains directional foils in addition. Look this up by googling it. A Francis turbine/pump design can also handle large volumes, tens of cubic meters per second. A Francis turbine has its limitations. When handling large pressures (water falls beyond a few hundred meters), the snail shell-shaped water distribution system needs to be extremely thick. Because of this, water is usually not pumped more than 200–300 meters. When water reservoirs and hydroelectric power stations with much larger water falls are built, one therefore usually applies a Pelton turbine. A Pelton turbine looks like a disk with several spoon-shaped cups attached at the edge. Water comes through several (often 4 or 5) nozzles and hits these cups making the turbine spin. The center of the turbine wheel is then attached to a generator by a shaft. A Pelton turbine wheel is shown in Fig. 3.3, left.

Two points worth bearing in mind are that i) when considering hydroelectric energy storage, water is usually not pumped to a too high elevation and ii) that for pumped hydroelectric energy storage, it is the Francis turbine that is the common technology. This does not mean that Pelton turbines and high water falls are not included in hydroelectric energy storage; this only means that Pelton turbines and high water reservoirs are not included in *pumped* hydroelectric energy source. It is actually so flexible that, in a larger mixed energy grid with e.g. solar and wind supply (which are intermittent ones), hydroelectric power stations with buffering capacity are considered energy storage units. In most countries where hydroelectric energy plays an important part of the electric energy production, water is therefore kept in large basins and only tapped onto the turbines when power is needed. Also, when the water comes from great heights, it has often arrived as snow during parts of the year and released during other parts of the year. This means that the energy is stored on the hill sides of lakes (dams or not) and then slowly accumulates into the lakes (basins) ready to be tapped on demand. This can be said to be the most important energy storage mechanism when it comes to hydroelectric energy storage.

Hydroelectric energy production and storage do not have to be very complicated, although one can study the field in great details and depth. Here we shall only look at the general governing equations. For any control volume (CV), we can use and expand the first law of thermodynamics (Eq. (2.4)) to define the energy conservation as in the equation

$$\frac{dE_{CV}}{dt} = \dot{Q} - \dot{W} + \dot{m}_{in}\left(U_{in} + \frac{v_{in}^2}{2} + gz_{in}\right) - \dot{m}_{out}\left(U_{out} + \frac{v_{out}^2}{2} + gz_{out}\right).$$

(3.15)

At stationary conditions, we have $\frac{dE}{dt} = 0$. The control volume work W_{CV}, which is transmitted to a shaft and further along with viscous work W_μ, and pressure drop for the volume stream is defined by the equation

$$\dot{W} = \dot{W}_\mu + \dot{W}_{CV} + \dot{V}\Delta p.$$

(3.16)

For most liquids, the heat capacity is so large that the fluid adsorbs the heat from friction without increasing significantly in temperature. Moreover, the liquid passes so quickly through the machines that it is actually very little time for the system to exchange heat $\dot{Q}$ with the surroundings. In other words, since the liquid hardly changes temperature from friction and because of very short residence times, heat is not exchanged with the surroundings, and $\dot{Q} \approx 0$. We consider incompressible flow, that is, constant volume. Since we consider the system to be isothermal ($\Delta T = 0$) and incompressible, the change in internal energy is also negligible ($\Delta U = C_v \Delta T$). Equation (3.15) turns into:

$$\dot{m}_{in}\left(\frac{v_{in}^2}{2} + gz_{in}\right) - \dot{m}_{out}\left(\frac{v_{out}^2}{2} + gz_{out}\right) = \dot{W}_{CV} + \dot{W}_\mu + \dot{V}\left(p_{out} - p_{in}\right),$$

(3.17)

$$\dot{m}_{in}\left(\frac{v_{in}^2}{2}+gz_{in}\right)-\dot{m}_{out}\left(\frac{v_{out}^2}{2}+gz_{out}\right)=\dot{W}_{CV}+\dot{W}_\mu+\dot{m}\left(\frac{\dot{V}}{\dot{m}}p_{out}-\frac{\dot{V}}{\dot{m}}p_{in}\right),$$

(3.18)

which can be rewritten into

$$\dot{m}_{in}\left(\frac{\dot{V}}{\dot{m}}p_{in}+\frac{v_{in}^2}{2}+gz_{in}\right)-\dot{m}_{out}\left(\frac{\dot{V}}{\dot{m}}p_{out}+\frac{v_{out}^2}{2}+gz_{out}\right)=\dot{W}_{CV}+\dot{W}_\mu$$

(3.19)

and in turn into

$$\left(\frac{p_{in}}{\rho g}+\frac{v_{in}^2}{2g}+z_{in}\right)-\left(\frac{p_{out}}{\rho g}+\frac{v_{out}^2}{2g}+z_{out}\right)=\frac{\dot{W}_{CV}}{\dot{m}g}+\frac{\dot{W}_\mu}{\dot{m}g}.$$
(3.20)

All the terms in Eq. (3.20) are represented by the units of meters and are generally known as heads or pressure heads. A pressure head refers to the pressure of water column of a certain height or head. When it comes to potential work, we can use the same term, head, to describe the potential work from a given mass of water at a given height (mass and the gravitational constant) and its potential work release in a control volume like a turbine. This potential work is, in the head term analogy, named turbine head, $h_{turbine}$. Likewise when pumping, the work that must be applied to the pumping (turbine) wheel now becomes h_{pump}. In any real system, there is also friction, and for hydroelectric ones, this friction is related to viscosity. The potential work that is lost to friction results in a lowering of the out head and is given the label h_μ or the viscous head. This head is considered a loss regardless of pumping or power extraction. If one is pumping water, the head of pumping, h_{pump}, is a flow inducing head and thus has the opposite sign of the viscous head. By dividing the work terms of the control volume by the weight ($\dot{m}g$) thus becomes

$$\frac{\dot{W}_{CV}}{\dot{m}g}+\frac{\dot{W}_\mu}{\dot{m}g}=h_{turbine}-h_{pump}+h_\mu.$$

When more generically calling entrance point 1 and exit point 2, by insertion into Eq. (3.20) we obtain:

$$\left(\frac{p}{\rho g}+\frac{v^2}{2g}+z\right)_{in}-\left(\frac{p}{\rho g}+\frac{v^2}{2g}+z\right)_{out}=h_\mu-h_{pump}+h_{turbine}.$$
(3.21)

Equation (3.21) is meant to be used on turbine and pumping systems. When it is applied for pumping, the turbine head is considered zero

($h_{turbine} = 0$) and vice versa when loading power the pumping head is considered zero ($h_{pump} = 0$). Therefore one of the two will always be zero when applied to a problem. Considering pumped hydro, the atmospheric pressure is usually the same at the inlet and outlet, so that the two pressures cancel each other. An exception is the lately envisioned submarine hydroelectric energy storage tanks exemplified in Example 3.4. Moreover, for hydroelectric energy systems, the water velocity is usually zero both at the inlet and outlet, and these terms cancels too. The velocities are zero because the control volume is in contact with water bodies at rest. If the inlet stream had velocity, then it would look like a funnel at the bottom of a water fall, in turn connected to a water gate and then to a turbine. If the outlet stream had a velocity, it would look like a fountain. The former is impractical because it would lead to lots of air bubbles in the turbine system, and the latter would represent an unnecessary energy dissipation. The point is that Eq. (3.21) for charging and discharging hydroelectric energy storage in most cases can be simplified to $z_1 - z_2 = h_\mu - h_{pump}$ and $z_1 - z_2 = h_\mu + h_{turbine}$, respectively.

Example 3.3: Efficiency of Hydroelectric Energy Storage.

Several water turbines are put on display like in Fig. 3.3. Next to them, one can often find some fact sheets. For the two turbines of Fig. 3.3, the operation facts are as follows:

	Power/MW	ω/rpm	Water fall/m	$\dot{V}$/m^3 s^{-1}	Radius/m
Pelton	56	428	740	8.8	1.6
Francis	50	130	300	51	1.3

These are actually quite impressive facts when one think about them the right way. Lets consider a few examples.

a) Give an estimate on how much time in seconds would it take to fill your bedroom for each of the volume streams $\dot{V}$?

b) What are the maximum powers for each of the turbines based on these volume streams?

c) What is the efficiency of the Francis turbine system?

d) What are the friction head h_{fric} and the turbine head $h_{turbine}$ for the Francis turbine considering that all energy lost is in friction, the water reservoirs in and out are at rest, and that both reservoirs are in contact with the nearby air?

e) What is reasonable overall efficiency if one pumps at half the rate as one extracts power back using the Francis turbine for pumped hydroelectric energy storage?

Solutions:

a) *Considering a student dormitory of 10 m^2 and 2.5 m up to the sealing a volume of 25 m^3 is reasonable. By dividing the volume by the volumetric flow rate we get the filling time. The filling time for the Pelton and the Francis turbine water streams are then (25/8.8 $\approx$) 3 seconds and (25/51 $\approx$) 0.5 seconds. These are impressive numbers! Realize first that your bedroom can be filled with water in 3 seconds and then next that it could be filled twice every second! That is what these turbines handle.*

b) *The maximum power is given by the mass flow, the gravitational constant, and the water fall height $h_{w.f.}$:*

$$P^{max} = \dot{m}gh_{w.f.} = \rho\dot{V}g(z_{in} - z_{out}).$$

For the two turbines, we thus get:

$$P^{max}_{Pelton} = 998\left[\frac{kg}{m^3}\right]8.8\left[\frac{m^3}{s}\right]9.81\left[\frac{m}{s^2}\right]740\,[m] = 63.8\,[MW],$$

$$P^{max}_F = 998\left[\frac{kg}{m^3}\right]51\left[\frac{m^3}{s}\right]9.81\left[\frac{m}{s^2}\right]150\,[m] = 74.9\,[MW].$$

c) *The efficiency ε is the ratio between the maximum and the actual power:*

$$\varepsilon = \frac{P}{P^{max}} = \frac{50\,[MW]}{74.9\,[MW]} = 67.1\%.$$

d) *Differentiating the friction head h_{fric} from the turbine head $h_{turbine}$ requires the use of Eq. (3.21). Because the water reservoirs are at rest, there is no kinetic energy involved, so that $v_{in} = v_{out} = 0$. Because the water reservoirs are in contact with the same air, there is no (significant) pressure differences between the inlet and outlet, that is, $p_{in} = p_{out}$. The water fall height $h_{w.f.}$ equals the difference between the inlet and outlet heights ($z_{in} - z_{out}$) regardless of where the turbines are in the pipe line system. Equation (3.21) thus becomes $h_{w.f.} = h_{fric} + h_{turbine}$.*
Still we are left with one equation and two unknowns. From the previous problem we see that the turbine height is the water fall height times the efficiency: $h_{turbine} = \varepsilon h_{w.f.} = 0.671 \cdot 150\,[m] = 100\,[m]$. This means that the turbine only experiences a hydraulic head equivalent to a water column of 100 m!
We can now calculate the friction head/height: $h_{fric} = h_{w.f.} - h_{turbine} = 150\,[m] - 100\,[m] = 50\,[m]$.

e) *Very often, water is pumped slower than power is extracted during peak hours. This is similar to most energy storage devices. See for instance the battery in Example 1.1. For laminar flow, the flow rate is proportional to the pressure drop, which in turn is proportional to the friction head h_{fric}. This means that since one is pumping at half the rate the pumping friction head is half the power friction* *head. The single efficiency is given by Eq. (1.2), and the overall efficiency for any process of several steps is given by Eq. (1.3). Thus we have:*

$$\varepsilon_{tot} = \varepsilon_{pump}\varepsilon_{power} = \frac{E_{pot}}{E_{pump}}\frac{E_{power}}{E_{pot}} = \frac{E_{power}}{E_{pump}} = \frac{P_{power}\Delta t}{P_{pump}2\Delta t}$$

$$= \frac{\rho g \dot{V} h_{turbine}\Delta t}{\rho g 0.5 \dot{V} h_{pump}2\Delta t} = \frac{h_{turbine}}{h_{pump}} = \frac{h_{w.f} - h_{fric,power}}{h_{w.f.} + |h_{fric,pump}|}$$

$$= \frac{h_{w.f} - h_{fric,power}}{h_{w.f.} + 0.5 h_{fric,power}} = \frac{150\,[\text{m}] - 50\,[\text{m}]}{150\,[\text{m}] + 25\,[\text{m}]} = \frac{100\,[\text{m}]}{175\,[\text{m}]} = 57.1\%.$$

The overall efficiency thus becomes 57%. Note that lowering the friction head is beneficial, but that also the water column is important.

Example 3.4: Hydroelectric Energy with $p_1 \neq p_2$.

One way to store energy using hydroelectric pumping is by inserting a large submarine rigid tank at the bottom of a sea bed. This tank then has a snorkel above the sea level, and in the bottom, it has a revertible turbine/pumping device, like, for instance, a Francis turbine. When pumping out the water from the tank in the bottom of the tank, a pressure difference gradually changes due to the lowering of the ambient hydrostatic pressure. The water does not change its gravimetric potential.

How does Eq. (3.21) appear for pumping and power extraction in this instance?

As an engineer, what do you consider as the biggest challenges with this system?

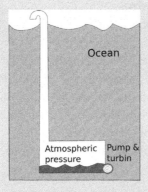

Since the water is at rest on both sides of the walls of the submarine tank, the velocity head ($\frac{v^2}{2g}$) cancels out, and due to no difference in elevation for the water inlet and outlet, the height heads also cancels. We are then left with $\frac{p_1}{\rho g} - \frac{p_2}{\rho g} = h_\mu - h_{pump}$ and $\frac{p_1}{\rho g} - \frac{p_2}{\rho g} = h_\mu + h_{turbine}$ for pumping and power extraction, respectively.

The challenges with this system are several. If we did not consider maintenance and corrosion issues, we could look at the volumetric size of this device. If we wanted to do something like in Example 3.3, 56 MW for an hour, we would need 55 m³ s⁻¹ for 3600 s, that is, 183600 m³. If we wanted to keep this within one bar pressure change, 10 m height in the submarine container, we would need $135 \times 135 \times 10$ meter underwater construction at 300 m depth at the sea bed. Moreover, this would also require a lot of concrete (or other mass) to avoid buoyancy problems. Since concrete has a density around 2.5 times that of water, we would need 73400 m³ of concrete. This is equivalent to 1.7 m-thick walls.

PROBLEMS

Problem 3.1. Flywheels: Size and Materials.

Consider a flywheel that is disk shaped with a height of 5 cm and a diameter of 20 cm. Calculate the energy and specific energy for the two given cases:

a) The rotation speed is 16 krpm, and the flywheel is made of steel with density 7.8 kg/L.
b) The rotation speed is allowed to be six times that of steel, and the density is a quarter.
c) What is the specific power for the two flywheels if the energy was released in ten seconds? Do not account for the engine weight.

Problem 3.2. Pumped Hydro.

A water reservoir is dug out on the top of a mountain. A river floats alongside the foot of the mountain, 500 meter below. At night, water is slowly pumped up from the river so that the pipe friction head loss is a third of what it is in the morning when the energy is released. Pumping results in a friction head loss of 10 m, and the turbine friction head loss in the morning is 40 m. The pipe friction head loss when discharging is twenty percent larger than the friction head of the turbine.

a) Based on Eq. (3.20), explain that here we can use the following expression:

$$z_{in} = z_{out} + h_{turb} + h_{\mu} - h_{pumpe}.$$

b) What is the efficiency for the entire cycle?
c) What is the major contribution to the loss in efficiency and how large is its portion?

SOLUTIONS

Solution to Problem 3.1. Flywheels: Velocity and Materials.

For both problems, we need Eq. (3.10)a to find the energy:

$$E_{kin} = \frac{\pi^3 \rho h \omega^2}{60^2} r^4.$$

Next, we find the mass: $m = \rho V = \rho h \pi r^2$.

a) For the steel wheel, we get:

$$E_{kin} = \frac{\pi^3 7800 \ [\text{kg/m}^3] 0.05 \ [\text{m}] 16000^2 \ [\text{rpm}^2]}{60^2 \ [\text{s}^2/\text{min}^2]} 0.1^4 \ [\text{m}]$$

$$= 86.0 \ [\text{kJ}] = 23.9 \ [\text{Wh}],$$

$$m = 7800 \ [\text{kg/m}^3] 0.05 \ [\text{m}] \pi 0.1^2 \ [\text{m}^2] = 12.2 \ [\text{kg}]$$

and, in turn,

$$e = \frac{23.8 \ [\text{Wh}]}{12.2 \ [\text{kg}]} = 1.95 \ [\text{Wh/kg}].$$

b) For the new wheel, we get:

$$E_{kin} = \frac{\pi^3 1950 \ [\text{kg/m}^3] 0.05 \ [\text{m}] 96000^2 \ [\text{rpm}^2]}{60^2 \ [\text{s}^2/\text{min}^2]} 0.1^4 \ [\text{m}]$$

$$= 774 \ [\text{kJ}] = 215 \ [\text{Wh}],$$

$$m = 1950 \ [\text{kg/m}^3] 0.05 \ [\text{m}] \pi 0.1^2 \ [\text{m}^2] = 3.06 \ [\text{kg}]$$

and, in turn,

$$e = \frac{215 \ [\text{Wh}]}{3.06 \ [\text{kg}]} = 70.3 \ [\text{Wh/kg}].$$

This is characteristic for composite materials relative to steel that they are lighter but tolerate more rotation velocity.

c) If this energy were converted in ten seconds, then the specific powers would be 0.70 and 25.3 kW/kg, which in the light of the Ragone chart of Chapter 1 is quite large. This is as the calculation does not account for the motor and other auxiliary components. Imagine a hybrid system where the flywheel would utilise an existing motor of a vehicle, then a system power density of the flywheel could potentially be in the order of 10 kW/kg.

Solution to Problem 3.2. Pumped Hydro.

a)

$$\left(\frac{p_{in}}{\rho g} + \frac{v_{in}^2}{2g} + z_{in}\right) - \left(\frac{p_{out}}{\rho g} + \frac{v_{out}^2}{2g} + z_{out}\right) = \frac{\dot{W}_{CV}}{\dot{m}g} + \frac{\dot{W}_\mu}{\dot{m}g},$$

where p is the atmospheric pressure, which is equal at both intake and outlet and thus cancel out, v is the velocity in each of the reservoirs, which are close to at rest, and thus also cancels out. It is at the top during discharge and at the bottom during pumping. In turn, this gives

$$z_{in} = z_{out} + h_{turb} + h_\mu - h_{pumpe}.$$

b)

	Pipe friction head	Wheel friction head	ε_i
Discharge	48 m	40 m	$\frac{500-88}{500} = 0.82$
Pumping	16 m	10 m	$\frac{500}{500+26} = 0.95$

The overall efficiency becomes the product of the two single ones: $0.82 \cdot 0.95 = 78\%$.

c) The most important contribution is the pipe friction during discharge. The relative contribution is $\frac{48}{48+40+10+16} = 42\%$.

CHAPTER 4

Thermal Energy Storage

In daily life, we store thermal energy or the deficit of thermal energy. The former can be exemplified by the heat in distributed heating network during night time, reusable heating pads, solar heat collection, and many other examples. The latter is best exemplified by night-time cooling of buildings, refrigeration of various food products, or storing snow from one winter to the next one. In this chapter, we look at the general thermodynamics of the systems that store a surplus of heat and what knowledge is needed to engineer these systems. If wanting to store deficit of heat, the problems are very similar, however, not discussed here.

Generally, the primary key to understand this chapter is that heat as a process quantity property equals the enthalpy change:

$$Q = \Delta H = H_{out} - H_{in}. \tag{4.1}$$

This very simple equation is derived from (2.21) (p. 23). In storing thermal heat, we apply the simplification that no work is included in the process and that the systems are open and undergo a change at constant pressure. In a (mass)-specific form, we obtain:

$$q = \sum_{i=1}^{n} h_{i,out} - \sum_{i=1}^{n} h_{i,in}. \tag{4.2}$$

The key point in all cases in this chapter is to determine the change in enthalpy. We further evaluate how to do this in various ways.

The secondary key in evaluating thermal energy storage is about quality rather than quantity. The benchmark for this evaluation relates to what the heat can be used for. Can we use heat for thermal engines? Is it better to use this heat for work than just heat? There is a quantitative answer to this qualitative question. It is given by the second law of thermodynamics and the Carnot efficiency (Section 2.2.2):

$$\frac{W_{max\ out}}{Q_{in}} = \varepsilon_{Carnot} = 1 - \frac{T_C}{T_H}. \tag{4.3}$$

In essence, we evaluate the quality of heat based on its ability to deliver useful work. But in a radiator at home, higher temperature heat also trans-

Engineering Energy Storage.
DOI: 10.1016/B978-0-12-814100-7.00004-3
Copyright © 2017 Elsevier Inc. All rights reserved.

mits better, so that higher temperature is also more practical. Now, in this chapter, we evaluate both quality and quantity of heat.

4.1 HEAT VS. THERMAL ENERGY

Among many engineers and scientists, the term heat has different meanings. Some are more picky, and some are pragmatic. Regardless of this, understanding their point of view is important when one sells an energy system on behalf of a company, discusses heat with the common man on the street, or works with hard core detailed obsessed academics. When we feel heat, it is the process of temperature increase at some point in our body. Thermal energy is transferred to our body, the temperature subsequently rises, and our nervous system makes us aware of it. On a daily basis, we say that heat is transferred. A thermodynamicist would claim that heat is transference of thermal energy, so that the term transport of heat is redundant. Heat is transport (of thermal energy), and thus mentioning transporting or transport process is just wrong. Moreover, the detailed obsessive thermal scientist might also argue that heat is not stored, but rather thermal energy can be absorbed by certain materials, leading either to rise in temperature, phase change, or endothermic reactions.

In this chapter and book, we get over it. Heat is thermal energy, and they can both be transported and stored. Endothermic processes requires heat, and exothermic processes delivers heat. Heat being delivered to a system has a positive value.

4.2 SENSIBLE HEAT

Sensible heat is a term commonly used for the heat release relative to a change in temperature. The term is well exemplified by liquid water, where the temperature increases close to proportionally the amount of heat transferred to the water. When stating that the heat is added, it is underlying that the enthalpy is changing accordingly. The difference in temperature and its relation to heat and enthalpy are given by the definition of heat capacity at constant pressure:

$$c_p = \left(\frac{dh}{dT} \right)_p \Leftrightarrow c_p \Delta T = \Delta h = \frac{\dot{Q}}{\dot{m}} \Rightarrow Q = mc_p \Delta T, \qquad (4.4)$$

where c_p is the specific heat capacity.

Example 4.1: Thermal Energy Capacity of District Heating.

The district heating system of a given city has a pipeline network where (for simplicity) all the pipelines have a diameter of 15 cm. The total pipeline network is 150-km long. The heat in the pipeline comes from the local waste incinerator that has a capacity of 60 MW. One morning, the temperature drops from 90 to 74°C between 6 pm and 8:30 am.

a) How much heat (in Wh and J) is associated with the temperature drop?

b) How much heat (in Wh and J) is delivered from the incinerator?

c) How much heat (in MW) is drawn on average by the consumers if considering a network loss of 12 MW?

Solution:

a) *Let us consider that all the water undergoes the same cycle and that the temperature of the water basin of the tubes drop 16°C. The heat of this cooling process (of the pipeline water) is given by the change in enthalpy as in Eq. (4.4):*

$$Q_{cooling} = m\Delta h = \rho_w V c_{p,w} \Delta T = \rho_w L\pi r^2 c_{p,w} \Delta T$$
$$= 998\,[\text{kg/m}^3]150\,[\text{km}]\pi 0.075^2\,[\text{m}^2]4.184\,[\text{kJ/(kg K)}]16\,[\text{K}]$$
$$= 177\,[\text{GJ}] = 49.2\,[\text{MWh}].$$

b) *The incinerator delivers heat rate integrated by time:*

$$Q_{incin.} = \dot{Q}\Delta t = 60\,[\text{MW}]2.5\,[\text{h}] = 150\,[\text{MWh}] = 540\,[\text{GJ}].$$

c) *The heat rate drawn by the population in this time frame requires a heat balance for stationary heat rates:*

$$0 = \dot{Q}_{incin.} + \dot{Q}_{drawn} + \dot{Q}_{cooling} + \dot{Q}_{lost},$$
$$\dot{Q}_{drawn} = -\dot{Q}_{incin.} - \dot{Q}_{cooling} - \dot{Q}_{lost},$$
$$\dot{Q}_{drawn} = -60\,[\text{MW}] - \frac{49.2\,[\text{MWh}]}{2.5\,[h]} + 12\,[\text{MW}] = -67.7\,[\text{MW}].$$

Comment: Because of the buffer that the heated water mass represents, district heating can be said to be thermal energy storage. The other thing is that the continuous heat loss in the pipe line means that it takes time to rebuild the heat in the pipeline. The capacity of the distributed heating system is only 48 MW when accounting for the heat lost from the network.

4.3 LATENT HEAT

When ice melts into water, the temperature remains at the melting temperature T_m for a long while and until all the ice has melted. Likewise during freezing, the body of water remains at the freezing (or melting) temperature for a long time while giving off heat. This heat, which is associated with the phase transitions, is often referred to as latent heat and can be recognized as the enthalpy of fusion ΔH_{fus}. The enthalpy of fusion is given by the difference between the enthalpy in the solid state and the liquid state at the melting temperature, h_{s,T_m} and h_{l,T_m}, respectively. Thus the heat associated with melting becomes

$$Q = m\Delta h_{fus} = m\left(h_{l,T_m} - h_{s,T_m}\right). \tag{4.5}$$

Melting ice is not really relevant for storing thermal energy. It is relevant for remote cooling of foods, beverages, or possibly injuries and bruises. The face change heat or latent heat in phase changes is, however, very interesting for storing thermal energy in small-scale distributed systems. The most relevant example is possibly paraffin wax in relation to solar heat collectors. A solar heat collector for residential buildings typically consists of radiant heat adsorbing surfaces on a copper film inside a vacuum tube. The copper is in turn connected to a heat transfer fluid (HTF), which brings heat to a thermal reservoir inside the house. Typical operating temperatures for such fluids is 50–70°C, meaning that the fluid enters the collector at 50°C and leaves at 70°C. The HTF is then recooled to 50°C in a heat exchanger, for example, a hot water tank. This hot water tank is then used for showers and cleaning or any other need for hot water. Using hot water to store, this heat has at least to disadvantages: i) When tapping heat later on, the temperature of the stored heat quickly falls, meaning that the quality (Carnot value) lowers. ii) During heating, the water becomes hotter, and the temperature difference needed to drive heat into the water (in the solar collector) is lowered. Recall that to transfer heat into a flowing fluid is (by Newton's heat law) proportional to the temperature difference between the fluid and the surface delivering the heat. Having a large tank of paraffin wax that melts around 50°C, we resolve these two problems and, additionally, need less volume.

Example 4.2: Paraffin Wax for Heat Storage.

We will compare heat storage by phase change or latent heat to heat storage of liquid water or sensible heat. Consider a tank that contains 80 L of the heat storage medium (water or paraffin wax). At the beginning of the day, the temperature of the tank is 40°C, and when getting to 66°C, the heat storage medium no longer adsorbs more heat because the temperature driving potential of HTF is too low

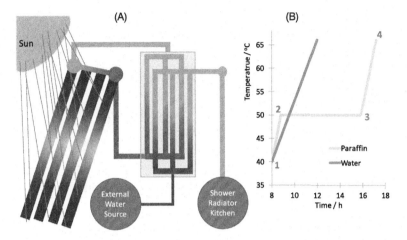

Figure 4.1 (A) One possible setup for heat storage and temperature behavior during heating using water (blue (dark gray in print versions)) or paraffin (yellow (light gray in print versions)) in Example 4.1 and (B) the time temperature behavior for the two 80 L systems.

(2°C in the tank and in the heat collector). Consider the following thermodynamic data:

$T_m/°C$	$c_{p,PW}/\frac{kJ}{kg\,K}$	$\Delta h_{PW}/\frac{kJ}{kg}$	$\rho_{PW}/\frac{kg}{m^3}$	$c_{p,w}/\frac{kJ}{kg\,K}$	$\rho_w/\frac{kg}{m^3}$
50	2.5	210	900	4.184	998

a) Sketch a possible system that includes the solar heaters in connection with the heat storage tank and water heating.
b) Calculate the volumetric thermodynamic data.
c) How much heat (in kJ and kWh) can 80 L of heat storage body (water and paraffin) take up during heating?
d) Make a graphical presentation of the time – temperature dependency for the heat storage body when 0.60 kW (2 m², 3 kWh day^{-1} m^{-2}, for 10 h day^{-1}) is continuously added.

Solution:
a) *See Fig. 4.1A. The solar heat is collected and transferred to the reservoir by the hot HTF (red (mid gray in print versions)). The reservoir consists of a tank of paraffin wax (yellow (light gray in print versions)) or water. The tank of heat storage medium has tubings (heat exchangers) separately for the HTF and the external water (blue (dark gray in print versions)) that is to be heated for use at home.*

b)

$\hat{c}_{p,PW}/\frac{kJ}{m^3\,K}$	$\Delta\hat{h}_{PW}/\frac{kJ}{m^3}$	$\hat{c}_{p,w}/\frac{kJ}{kg\,K}$
2250	189 000	4176

c) *Water: the heat uptake for water is*

$$Q = \Delta H = V \hat{c}_{p,w} \Delta T$$

$$= 0.08 \, [\text{m}^3] 4176 \left[\frac{\text{kJ}}{\text{m}^3 \text{K}} \right] 26 \, [\text{K}] = 8685 \, [\text{kJ}] = 2.41 \, [\text{kWh}].$$

Paraffin wax: the heat uptake comes in three portions: heating the solid (1–2), melting (2–3), and then heating the liquid wax (3–4)

$$Q_{1-2} = \Delta H = V \hat{c}_{p,PW} \Delta T$$

$$= 0.08 \, [\text{m}^3] 2250 \left[\frac{\text{kJ}}{\text{m}^3 \text{K}} \right] 10 \, [\text{K}] = 1800 \, [\text{kJ}] = 0.500 \, [\text{kWh}],$$

$$Q_{2-3} = \Delta H = V \hat{h}_{fus,PW} \Delta T$$

$$= 0.08 \, [\text{m}^3] 4176 \left[\frac{\text{kJ}}{\text{m}^3} \right] = 15120 \, [\text{kJ}] = 4.20 \, [\text{kWh}],$$

$$Q_{3-4} = \Delta H = V \hat{c}_{p,PW} \Delta T$$

$$= 0.08 \, [\text{m}^3] 2250 \left[\frac{\text{kJ}}{\text{m}^3 \text{K}} \right] 16 \, [\text{K}] = 2880 \, [\text{kJ}] = 0.800 \, [\text{kWh}].$$

In total: $Q_{tot} = 19\,800 \, [kJ] = 5.50 \, [kWh]$.
Comment: The paraffin wax stores about twice as much heat as the water for the selected temperature range.
d)   *See Fig. 4.1.*

4.4 REACTION HEAT

Two components reacting will lead to heat being released or adsorbed. For a chemical reaction going, this heat is equal to the reaction enthalpy ΔH. Consider any arbitrary reaction, here illustrated using one A and two B reacting to AB_2, then we can calculate the reaction heat from the molar enthalpy change:

$$A + 2B \leftrightharpoons AB_2, \tag{4.6}$$

$$Q = \Delta H = m_{AB_2} M_{AB_2} \left(\bar{h}_{AB_2} - \bar{h}_A - 2\bar{h}_B \right), \tag{4.7}$$

where M_i is the molar mass. This way, it is fairly easy to evaluate the chemical reaction heat by a series of reactions of this kind. Typically, this is hydrates that form at some temperature. The heat is released during reaction, and by adding heat to the system at sufficiently high temperatures we recharge the system. The heat can then be released by an initiator, like the coin-like thing in heating pads. Examples of some possible systems are given in Table 4.1.

There are two very important limiting factors that must be understood for this type of systems. The first is that one must go to or above a certain temperature to recharge the system. The other is that, unlike to the phase change materials in Section 4.4, one must convert all the substance back to the initial form (here $A + 2B$) before the recharging process is done. The first limitation is similar to that of phase change materials and a melting process, as the transition temperature must be reached, but the second one is different. In the heat reservoir of Example 4.2, we could melt, for example, 99% of the paraffin wax and leave the system at rest until heat is needed. (The heat reservoir will leak heat and inevitable and slowly discharge the phase change heat.) In the case of reaction heat like that in Eq. (4.6), converting 99% and trying to leave the system at rest would lead to the entire heat added immediately being released. This may sound strange and inconvenient, but remember that once the entire system is charged in astable manner, the reaction heat system can be stored for weeks and years and still being available when needed. This is very convenient, as long as the charging reaction completed. The strangeness about the reaction having to go all the way is elaborated in the next paragraph.

Understanding the heat storage kinetics and the need for initial heat triggered at will require one more piece of thermodynamics. What drives a reaction is chemical potential, or Gibbs free energy (see Section 2.5). We can define molar Gibbs free energy for the mixture of A and 2B, $\bar{g}_{A+2B}$, and we can define the molar Gibbs free energy for AB$_2$, $\bar{g}_{AB_2}$. The molar Gibbs free energy then becomes

$$\Delta_f \bar{g} = \bar{g}_{AB_2} - \bar{g}_A - 2\bar{g}_B \tag{4.8}$$

when the Gibbs free energy of formation $\Delta_f \bar{g}$ becomes negative, and the reaction is spontaneous and will go by itself. The only thing needed is the activation energy $\Delta_{act}\bar{g}$. The activation energy represents a barrier, which the reactants must overcome in order for the reaction to start. Once this barrier is overcome, the energy gain will be large, and the net Gibbs free energy of the reaction $\Delta_f \bar{g}$ will be negative. These energy levels are indicated in Fig. 4.2, where we can see that the chemical composition of A and 2B is kinetically trapped and cannot be released until overcoming the activation energy. Once the reaction goes, all of the reaction takes place, and the entire amount of A and B reacts into AB$_2$. There are many examples of reactions like this, that is, reactions where the reactants are stored at

Table 4.1 Thermodynamic data for some reaction heat of aqueous-based complex formers and relevant transition temperatures

Reaction	Δh^o / kJ mole^{-1}	q^o / kJ kg^{-1}	$T_{tr,\Delta\bar{g}^o=0}$ / °C	$q_{act.}$ / kJ kg^{-1}	$T_{tr,\Delta\bar{g}=0}$ / °C	K_{tr}
$NaC_2H_3O_2 + 3\,H_2O \leftrightharpoons NaC_2H_3O_2 \cdot 3\,H_2O$	−36	265	156	∼ 280[a]	58[a]	$5 \cdot 10^{-2}$[a]
$Ba(OH)_2 + 8\,H_2O \leftrightharpoons Ba(OH)_2 \cdot 8(H_2O)$	−109	346	205	266	78	$5 \cdot 10^{-5}$
$Na_2S_2O_3 + 3H_2O \leftrightharpoons Na_2S_2O_3 \cdot 3H_2O$	−55	222	158	203	48	$5 \cdot 10^{-3}$
$Mg(NO_3)_2 + 6\,H_2O \leftrightharpoons Mg(NO_3)_2 \cdot 6\,H_2O$	−106	414	581	163	89	$2 \cdot 10^{-8}$
$MgCl_2 + 6\,H_2O \leftrightharpoons MgCl_2 \cdot 6\,H_2O$	−142	700	759	169	117	$1 \cdot 10^{-20}$
References	[2]	[2]	[2]	[5]	[5]	

[a] Separate references: [6,7]

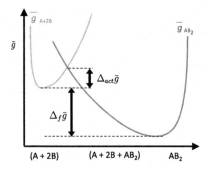

Figure 4.2 Energy stability curves for mixtures and compositions relevant for heat storages via reaction heat.

rest until initiated. Gasoline and air constitute one example where it take very little energy to initiate the energy release, and nitroglycerin is another. With the reactions in this section, the systems are regenerative by adding heat. Some examples are given in Table 4.1.

The condition to recharge the system, that is, going from AB_2 to A and 2B, requires heating the system to a temperature where the chemical potential or Gibbs free energy for formation of AB_2 is positive. A simplified, however, incorrect, explanation of this is to evaluate the standard Gibbs free energy of the reaction, $\Delta_r \bar{g}^o$, as a function of temperature, Eq. (4.9), and when this value becomes zero, Eq. (4.10):

$$\Delta_f \bar{g}^o = \Delta_f \bar{h}^o - T \Delta_f \bar{s}^o, \tag{4.9}$$

$$\Delta_f \bar{g}^o = 0; \quad T_{tr.,\bar{g}^o=0} = \frac{\Delta_f \bar{h}^o}{\Delta_f \bar{s}^o}. \tag{4.10}$$

The problem with this formulation is that it considers the ratio between reactants and product more commonly known as the equilibrium constant K, to be unity (1.0), which is usually incorrect. When Gibbs free energy of the reaction, $\Delta_r \bar{g}$, changes sign, however, the ratio is several orders of magnitude different. The equilibrium constant or product reactant ratio is then

$$K_{tr.} = \frac{[AB_2]}{[A][B]^2} = \exp\left(\frac{\Delta \bar{h}^o - T_{tr.}\Delta \bar{s}^o}{\bar{R} T_{tr.}}\right), \tag{4.11}$$

which comes from Eq. (2.35): $\Delta \bar{g} = \Delta \bar{g}^o + \bar{R} T_{tr.} \ln K_{tr.} = 0$.

Looking at Table 4.1, these equilibrium constants at the actual transition temperatures (when $\Delta_r \bar{g} = 0$) are listed for some of the example relevant salt. Note that when forming complex hydrates, we denote it A·2B rather than AB_2; however the composition constant K remains the same. The equilibrium composition constant K in Table 4.1 gives the ratio between the amounts of crystals (A·2B) and the dissolved salt (A), although not directly. The constant tells us how many times more the product of the reactants we have compared to the product. This is explained better via Example 4.3.

In conclusion, to recharge a "reversible reaction heat thermal energy storage device," the mixture must be heated to the temperature where the Gibbs free energy of the reaction $\Delta_r \bar{g}$ is equal to zero. At this temperature the equilibrium is driven far to the left (of Eq. (4.6)), leading to an equilibrium constant several orders of magnitude smaller than unity. It is the reaction enthalpy ΔH that determines the available/required heat, and the standard enthalpy ΔH^o is only a first–order approximation.

Example 4.3: Enthalpy, Specific Heat, Transition Temperature and Equilibrium Constant.

Sodium acetate trihydrate $NaC_2H_3O_2 \cdot 3\,H_2O$ is the most common reversible reaction heat storage device, commonly used in heating pads available in many shops. Looking up the thermodynamic values in, for example, S I Chemical Data [2], determine

a) the standard enthalpy,
b) the standard specific heat,
c) the standard transition temperature,
d) the equilibrium constant when the actual transition temperature is $58°C$,
e) the ratio of the dissolved salt and the hydrate salt at $58°C$.
f) Why is the actual specific heat different from the standard one and how good approximation is the use of the standard enthalpy?

Solution:

a) *The relevant standard (formation) energies are found in a table, and accordingly the standard reaction entropy is determined alongside:*

	$\Delta_{f/r}\bar{h}^o/$ kJ mole^{-1}	$\Delta_{f/r}\bar{g}^o$ KJ mole^{-1}	$\Delta_r\bar{s}^o/$ kJ mole^{-1} K^{-1}
$NaC_2H_3O_2$	-709	-607	
H_2O	-286	-237	
$NaC_2H_3O_2 \cdot 3\,H_2O$	-1603	-1329	
$NaC_2H_3O_2 + 3\,H_2O \rightleftharpoons NaC_2H_3O_2 \cdot 3\,H_2O$	-36	-11	-0.0839

f/r denotes that either formation or reaction energy (entropy) is listed.

b) *The molar weight of the salt complex is 136 mol g^{-1}. The specific heat thus becomes*

$$q = \Delta_r \bar{h}^o M_{NaC_2H_3O_2 \cdot 3H_2O}$$
$$= -36 \text{ [kJ/mole]}/0.136 \text{ [kg/mole]} = -265 \text{ [kJ/kg]}.$$

c) *The standard state transition temperature is given by Eq. (4.10):*

$$T_{tr., \bar{g}^o = 0} = \frac{\Delta_f \bar{h}^o}{\Delta_f \bar{s}^o} = \frac{-36 \text{ [kJ/mole]}}{-0.0839 \text{ [kJ/mole K]}} = 429 \text{ K} = 156°\text{C}.$$

d) *the equilibrium constant for a given temperature is given by*

$$K_{tr.} = \frac{[NaC_2H_3O_2 \cdot 3H_2O]}{[NaC_2H_3O_2][H_2O]^3} = \exp\left(\frac{\Delta \bar{h}^o - T_{tr.} \Delta \bar{s}^o}{\bar{R} T_{tr.}}\right)$$
$$= \exp\left(\frac{-36\,000 - 331 \cdot (-84) \text{ [J/mole]}}{8.314 \cdot 331 \text{ [J/mole]}}\right) = 5.02 \cdot 10^{-2}.$$

e) *The ratio between the dissolved salt ($NaC_2H_3O_2$) and the hydrate ($NaC_2H_3O_2 \cdot 3H_2O$) is $\frac{NaC_2H_3O_2 \cdot 3H_2O}{NaC_2H_3O_2 \cdot 3H_2O}$. Since we have 3 water per sodium acetate, the equilibrium constant becomes the inverse ratio we are after to the power of four $(3 + 1)$. Thus we get $\frac{NaC_2H_3O_2 \cdot 3H_2O}{NaC_2H_3O_2 \cdot 3H_2O} = K^{-1/4} = 2.1$. This means that when the system is recharged, we have one ninth of the product hydrate, two ninth of the reactant salt, and two thirds (6/9) of water on a molar basis.*

f) *Because we do not have all the salt at standard condition, the standard enthalpy is not suited for determining the reaction enthalpy. In other words, the reaction heat and temperature must be found tabulated elsewhere.*

4.5 EUTECTIC AND NONEUTECTIC HEAT

So far we have evaluated heating and cooling of a single phase (4.2), a phase transition (4.3), and a reversible reaction heat at a given temperature (4.4). The processes have become increasingly complicated as we have progressed, and in this final section, we look at the combination of a phase being cooled while simultaneously having a phase transition or a reaction heat.

Again, we choose to present some arbitrary mixture of substances A and B, each with different melting temperatures $T_{m,A}$ and $T_{m,B}$. In this case, they solidify by forming the compound AB_2 and some A or B, depending on the residual amount of A or B beyond the stoichiometric A:2B amount. When the compound AB_2 is formed, the melting temperature is as low as it can get for this mixture. We denote this the eutectic melting tempera-

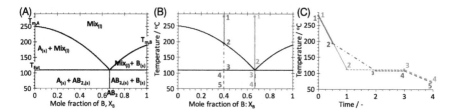

Figure 4.3 (A) Phase diagram for the binary mixture of A and B, solidifying into AB$_2$. (B) and (C): Cooling paths in time and compositions for two example mixtures.

ture T_{Eut} of AB$_2$ by T_{Eut,AB_2}. The process of crystallizing this mixture is best explained and visualized using a phase diagram. Since there are two components, we call this a *binary* phase diagram. The example binary phase diagram with its characteristics is shown in Fig. 4.3A, whereas the processes of cooling two different compositions (blue (mid gray in print versions) and orange (light gray in print versions)) are illustrated in Figs. 4.3B and C.

The eutectic mixture in this example consists of one third of A and two thirds of B, that is, $2n_A = n_B$, where n_i denotes amount of moles of component i. This mixture is the eutectic mixture and is characterized by that when cooling, its cooling behavior is much like that of the latent heat phase transition in Section 4.3. The cooling process of this mixture when heat leaves at a constant rate is indicated using orange (light gray in print versions) in the simplified phase diagram and the temperature time behavior in Figs. 4.3B and C. The eutectic mixture follows the orange (light gray in print versions) cooling line;

1–2: The liquid mixture consisting of 66.7 mol% B is cooled until the eutectic melting temperature T_{Eut} is met (solid line).

2–3: The entire mixture undergoes a phase change and solidifies into AB$_{2,(s)}$ (short dash). This process takes some time as the phase transition gives off heat at constant temperature.

3–4: Once the entire mixture is solidified, the temperature continues to cool further (long dash).

A noneutectic mixture is any mixture that has a composition different from the eutectic one, that is, $2n_A \neq n_B$. Because it is very difficult to exactly meet the criteria of a eutectic composition, eutectic freezing effects almost always happen to some extent just above the melting temperature; however, these are usually negligible. When shifting significantly away from the eutectic composition, the solidification process changes into a smelting process (melting is defined as the process of a single compound changing phase from solid to liquid, whereas smelting is defined as a process where different compounds melts or solidifies together and gradually change composition, like, e.g., a noneutectic one), and the temperature time behavior becomes more complex. One example of this (again for heat leaving at a constant rate) is illustrated using blue (mid gray in print versions) and a composition of 0.4 mol% B. Because this composition contains less amount of B than required for the eutectic mixture, this mixture is termed under eutectic.

1–2: The entire liquid mixture cools until meeting the liquidus line (solid line).

2–3: From this point and until the system is cooled to the eutectic temperature, solid A will precipitate (dash-dotted line). Because more heat is released from the phase transition of solidifying A, the cooling rate (temperature falling) becomes lower. This can be calculated from enthalpy of combined cooling partial eutectic mixture and solidifying partial amount of A.

3–4: Once the system has cooled to the eutectic temperature, the rest of the mixture (now solid A and the remaining liquid being "purified" into eutectic composition liquid mixture) solidifies (short dashed line).

4–5: The cooling as a single solid continues (long dashed line).

Likewise to this, the process would first solidify B for over eutectic compositions. The behaviour in this example requires that the molar enthalpy of fusion of A, B, and AB_2 have one same value and likewise for the molar heat capacities.

In this chapter, we have so far evaluated the phase change and that the heat associated with it is something valuable. If operating a cooling fluid that is a mixture, and especially a eutectic mixture, then one must take care make sure that the mixture is well mixed when meeting temperatures are close to the eutectic temperature. If the mixture is locally out of eutectic composition, which can naturally happen in reservoirs with high temperature differences, the mixture could cool likewise to the blue line and precipitate a solid and clog the cooling loop. The natural process that induces such local mixing deviations at different temperatures in a smelt is termed the Soret equilibrium; see [8] for further reading.

Ternary, quartiary, and higher-order phase diagrams also exist. A ternary mixture (A, B, and C) usually has a eutectic melting point that is lower than any of the eutectic melting point of the binary mixtures (AB, AC, or BC), and the melting temperature of the composition nearby the eutectic mixture is often more sensitive than that of the binary. This is important to consider when operating, for instance, solar power towers (see Section 5.3).

PROBLEMS

Problem 4.1. Heat Management in Fast Charge for Batteries.

During fast charging of Li-ion batteries, heat is developed and released. When fast charging, this is mainly due to ohmic heating RI^2. There are several reasons for why one does not want the battery significantly above 50°C.

a) How can we utilize paraffin wax (PW) as a phase transition material to prevent overheating in such an instance?

Consider that the battery subject to charging generates 10 W/kg of heat in addition to heating itself as it increases in temperature under this charging cycle. Once at the melting temperature of the PW, the heat delivered increases by a factor of three (as the battery is at constant temperature and no longer heat itself).

The goal of this exercise is to determine how much PW is needed to avoid overheating of the battery.

b) How much heat (Wh) is adsorbed per kg of PW? Consider the system to start at 25°C and the PW to melt at 50°C. The heat capacity and heat of fusion of PW is 2.5 kJ/kg K and 210 kJ/kg, respectively.

c) How much heat is given off per kg of battery in this process?

d) How many kg PW is then needed per kg battery for all the PW to melt (but not increase further in temperature)?

e) How large fraction of the adsorbed heat contributes to the phase transition of the PW?

f) Consider a battery being charged. However, the battery is already at the melting point of the PW (still all solid), and the charging time is doubled. How does this affect the need for PW?

SOLUTIONS

Solution to Problem 4.1. Heat Management in Fast Charge for Batteries.

a) Keeping the Li-ion battery cells in a body of PW will lower the rate of temperature increase and keep the batteries at the melting temperature until all the PW is melted.

b) When the temperature raises from 25 to 50°C, the specific heat absorption is

$$2.5 \, \frac{kJ}{kg \cdot K} \cdot 25 \, K = 62.5 \, \frac{kJ}{kg} = 17 \, \frac{Wh}{kg},$$

and at 50°C, melting 1 kg PW requires

$$210 \, \frac{kJ}{kg} = 58 \, \frac{Wh}{kg}.$$

In turn, this means that 1 kg of PW can take up

$$17 \, \frac{Wh}{kg} + 58 \, \frac{Wh}{kg} = 75 \, \frac{Wh}{kg_{PW}},$$

when going from 25°C to the melting point and then melt all the PW.

c) From the problem formulation we see that one must take up

$$10 \, \frac{Wh}{kg} + 3 \cdot 10 \, \frac{Wh}{kg} = 40 \, \frac{Wh}{kg_{batt}}.$$

d) We now need

$$\frac{40 \, \frac{Wh}{kg_{Batt.}}}{75 \, \frac{Wh}{kg_{PW}}} = 0.533 \, \frac{kg_{PW}}{kg_{Batt.}}.$$

e) By comparing the amount of specific heat of fusion to the total amount of specific heat we get

$$\frac{58 \, \frac{Wh}{kg}}{75 \, \frac{Wh}{kg}} = 77.3\%.$$

e) Doubling the charging time means lowering the current bay a factor of two. In turn the heat power will then lower by a factor of four (RI^2). Since the time is the double, the total amount of heat is reduced by a factor of two. Considering that more than three quarters of the initial heat were used for the melting process and that we now have half the heat, we need less PW.

CHAPTER 5

Thermomechanical Energy Storage

5.1 THERMODYNAMICS: HEAT, WORK, AND STATES

One can store energy in the form of mechanical work (Chapter 3), heat (Chapter 4), or chemical energy (Chapters 6–9). The interplay of these forms of energy, and in particular of heat, can be both benign and malignant. Taking advantage of heat that is otherwise not used is very advantageous, and this is possibly one of the strongest growing fields within energy research. Heat is a low-quality form of energy, so taking advantage of it is a lower hanging fruit compared to other higher forms of energy. In this chapter, we evaluate two forms of energy storage where the interplay between mechanical energy and thermal energy is a key factor. This is energy storage by compression of air and thermal energy storage in solar power towers.

5.2 COMPRESSED AIR ENERGY STORAGE

Energy storage by compression air seems initially as a simple and smart way to store energy. Think of a pump for your bike and how much force you have to apply to press the piston while sealing the outlet with your thumb. The pump also bounces quickly out again. Compressed air energy storage (CAES) requires a revertible air turbine, a revertible electric generator, and a large cavern for storing the compressed air. In principle, it is appears an efficient system and a great idea, however, not so when doing more thorough analysis.

Compressed air energy storage (CAES) is illustrated in Fig. 5.1. Electric energy is stored by compressing air in a compressor. The energy can come from a surplus in the electric market or also from some base load power production unit nearby like, for example, a coal power plant or natural gas (NG) power plant. The point is that some surplus cheap electricity is dumped into the electrically driven air compressor. Next, the compressed air is fed into a reservoir, which for practical reasons often is an old salt mine. (Salt in mines like these where once extracted by drilling a hole into

Engineering Energy Storage.
DOI: 10.1016/B978-0-12-814100-7.00005-5
Copyright © 2017 Elsevier Inc. All rights reserved.

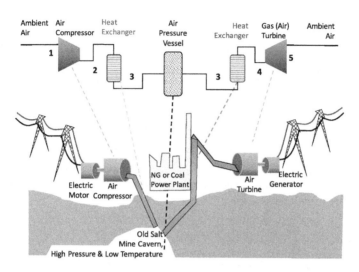

Figure 5.1 Gas process scheme (upper) for the compressed gas energy storage system (below). The system is based on first compressing air, second sending the air to a cavern where it is ultimately cooled, and third reheating the high-pressure gas, before finally depressurizing it in a gas (air) turbine.

the ground and down to the layer where salt rich minerals were located. Next, the salt was washed out using water, eventually leaving a cavern that can easily be air tight.) The compressed air gets heated by the isentropic compression and is then cooled while entering the high-pressure cavern. This is illustrated by the red (light gray in print versions) and blue (mid gray in print versions) gradients on the piping in Fig. 5.1. The energy is regenerated by discharging the compressed air in an adiabatic turbine. In this process, the air is cooled isentropically. To avoid freezing and the related impracticalities, the air is first reheated. This can be done by using waste heat in a nearby energy power plant with waste heat (coal, natural gas, or nuclear), fueled by, for instance, nuclear energy, NG, or coal. Thus the air goes through four process steps in this cycle; compression, cooling, heating, and expansion.

The process scheme of the CAES is illustrated in the upper part of Fig. 5.1. First (1–2) the air is compressed adiabatically. Considering this to be done in a reversibly, we can consider this process step isentropic and use the isentropic gas table for air compression in Table B (p. 209). For process equipment like this, we relate the necessary work to Eq. (4.2) (p. 47), considering no heat, kinetic, nor gravimetric energy contributions. Subsequently, after compression, the gas is cooled (2–3) while being deposited in the cavern. This is shown by the blue (mid gray in print versions) heat exchanger in Fig. 5.1. This process is isobaric, and the heat change at constant pressure is given by the change in enthalpy, also given in Table B. The compressed air is now at rest. Before generating electric energy from the compressed air, it is reheated. This is illustrated by a heat exchanger (3–4) in Fig. 5.1. The last step in a CAES cycle, is expanding the high pressure high temperature gas, (4–5) in Fig. 5.1. This is, like the compression step (1–2), a reversible and adiabtic (then isentropic) one that can be evaluated using the samee gas tables. Depending on whether the gas is heated higher or lower in temperature compared to the temperature after compression, the final temperature at atmospheric pressure is higher or lower than the ambien (intitial intake) temperature.

When evaluating the energy efficiency it is clear that using less heat to reheat gives better energy efficiency (more work per added heat), but also lower power output. Additionally, if the gas is cooled sufficiently during expansion (not sufficintly reheated upon expansion) water droplets and ice particles will form. These will wear down the turbine blades very quickly, which is obviously unwanted. If one can not heat the gas sufficintly upon expansion, the expansion could be done in two step with rehating in between. This requires an extra turbine and heat exchanger, and in turn more investment cost. The practical result is therefore a need for sufficiently reheating to avoid condensing phenomena. An example on how to evaluate enrgy in a CAES is given in Example 5.1.

Example 5.1: Compressed Air Energy Storage.

As explained, CAES is one way to store energy. It deals with both heat and work in four distinct process steps. Consider the following case:

A natural gas (NG) power station has the opportunity to compress air at night and store this compressed air in an old salt mine nearby. The ambient air is 7°C and 1 atm at night. The air in the old mine holds 29.4 atm and 32°C. Upon expansion during peak hours in the morning, the air is reheated to 377°C using exhaust gas, which does not really have any fiscal costs. Eventually, the heat is released at 1 atm.

a) Search the internet for compressed gas energy storage and sketch the process using traditional process units. Make a table that has a row for each of the relevant states, pressure p, temperature T, and relative pressure p_r according to Appendix B and Eq. (2.28), and enthalpy h.
b) Define all the specific enthalpies between each step.
c) Determine the specific heat and work of the four process steps.
d) What energy is regenerated per kg of air?

e) What are the efficiencies with respect to i) work out per work in, ii) work out per energy input, and iii) considering the heat and work for free. Comment the answers.

Solution:

a) *See Fig. 5.1.*

b) *The challenge in this type of problems is to first figure out the knowns. In the table below, all knowns given by the problem and the corresponding gas table properties are indicated by a bold font. In this instance, all thermodynamic data are available once the temperature is known. This is the case for the states 1, 3, and 4. The temperatures of states 2 and 5 are a little more cumbersome to determine. We must first determine the relative pressure and use this instead of temperature to find all the thermodynamic data. This is done by taking the former relative pressure and multiplying it with the pressure ration change (see Eq. (2.28), p. 24). The new relative pressure is shown by italics.*

	T/K	$T/°C$	p/atm	p_r	$h/kJ\,kg^{-1}$	Trick:
1	**280**	**3**	**1**	**0.804**	**280**	*Choose temperature in gas table*
2	740	447	29.4	*23.6*	735	$p_{r,2} = p_{r,1}\frac{p_2}{p_1} = 0.804\frac{29.4}{1} = 23.6$
3	**305**	**32**	**29.4**	–	**306**	*Choose temperature in gas table*
4	**650**	**377**	**29.4**	**16.1**	**660**	*Choose temperature in gas table*
5	~ 250	~ -25	**1**	*0.548*	245	$p_{r,5} = p_{r,4}\frac{p_5}{p_4} = 0.548$

c) *From Eq. (4.2) we get the specific work and heat for each of the steps:*

$w_{1-2} = h_1 - h_2 = -455\,[kJ\,kg^{-1}]$	$q_{2-3} = h_3 - h_2 = -429\,[kJ\,kg^{-1}]$
$q_{3-4} = h_4 - h_3 = 354\,[kJ\,kg^{-1}]$	$w_{4-5} = h_4 - h_5 = 415\,[kJ\,kg^{-1}]$

d) *We regenerate $415\,kJ\,kg_{air}^{-1}$.*

e) $\varepsilon_i = \frac{w_{out}}{w_{in}} = 91\%$, $\varepsilon_{ii} = \frac{w_{out}}{w_{in}+q_{in}} = 52\%$, *and $\varepsilon_{iii} = \infty$. The first option evaluates only electric work because this is what one usually pays for, though case ii) is what one would thermodynamically argue. Since this technology is based on utilizing electricity and heat available for free, the initial business case becomes infinitely good. However, one must of course pay for the initial installation.*

From Example 5.1 we see that the energy efficiency of a CAES system is really not the best. The main energy loss is losing the heat after compression and having to regenerate this. One can argue that the example accounts for too little reheating as the isentropic process ends at $-25°C$. Going to this low temperature can introduce brittleness in the steel of the turbine blades and formation of ice particles,which will erode the turbine and its blades. One should therefore heat the air to higher temperature or reheat the air in the expansion cycle.

On the positive side, CAES offers a solution to store very large amounts of energy in some instances in an economically reasonable way. When evaluating energy storage systems, one should bear in mind that because the need for energy storage in a renewable energy market is so vast, there will be a market for almost any type of storage technology. In a renewable context, a CAES system, in combination with a bio-fueled power plant, could give enormous gain. For example, one could compress air when energy supply is large and have better use of heat from a combustion process. There exist a few CAES systems already, and there are many opportunities of developing this technology

5.2.1 Cryogenic Energy Storage

An extension of CAES is liquid air energy storage (LAES). Instead of compressing air, the air is cooled and compressed in several steps to the point where it liquefies. The energy is restored by boiling the air at high pressure and expanding it in a turbine, similarly to the process in a Rankine cycle. Likewise to CAES, LEAS has very low energy efficiency, unless done in conjunction with residual (waste) heat from another process. Reuse and conservation of heat can be undertaken in at least two different ways. The first approach is that the boiling process of the liquid air takes place at low temperature and that this uses the heat from the surrounding in a local heat reservoir, established at a very low temperature. This low-temperature reservoir is filled with low-temperature heat when cooling and compressing the air upon liquefaction, and the heat is then recovered when preheating the air upon boiling. In a daily language, we should use the term cold storage, although this is thermodynamically incorrect. The approach for waste heat utilization is that low-grade waste heat can be used to superheat the compressed air before the air enters the turbine or a series of expansions in the temperature range of less than around 150°C.

5.2.2 Other Compressed Gases

Caverns can be used to compress any gas and store them there and to release the energy in expansion. One example is natural gas or methane. Natural gas and biogas are rich in methane and thus the sources. Storing these gases under high pressure in large old salt mines is a convenient way to store them in large quantities and for long times, meaning weeks or months. Natural gas has already a large buffer capacity in the existing grid networks. This means that it will continue to flow and distribute for days if the supply was to be cut off at the source. In turn, this means that although NG salt cavern storage is useful, it is an additional mean of energy storage to the intrinsic buffer capacity of the large NG grids.

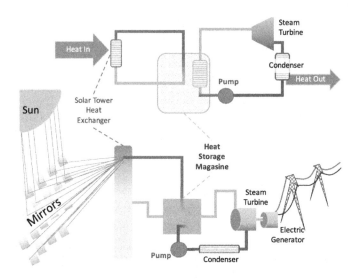

Figure 5.2 Process scheme (upper) for solar power towers as illustrated (lower). On day-time, sunlight is gathered at a focus point capturing heat and sending this heat to a reservoir (light purple (light gray in print versions)). The heat in this reservoir can be used to drive a Rankine cycle producing electric work.

5.3 SOLAR POWER TOWERS

Solar energy is most commonly associated with photovoltaic (PV) cells or solar cells. This technology ranges from 10 to 20%, depending on the PV materials and quality. The technology fits very well with small-scale instal-lations and is therefore so common in relation to buildings. For of-grid solutions, it needs a battery to supply energy when the sun no longer shines. Moreover, due to energy policies in certain countries, PV instal-lations dominate the electricity supply at certain hours, and for the most intense peaks, surplus electricity production has become an energy political challenge. Situations like this call for energy storage, and solar power towers offer a solution!

The principle of a solar power tower system is illustrated in Fig. 5.2. Large amounts of reflectors are placed around the solar tower. The mirrors focus the sunlight into one area of the tower, thus gathering all the sunlight into this spot. Obviously, this spot becomes very hot, and the heat from the sunlight is then used to heat a fluid, typically a molten salt. This molten salt is then returned to a reservoir where a lot of this salt is gathered in a closed basin. During daytime, the temperature of this salt increases. In other words, the heat is stored in a pool of molten salt available whenever needed.

The heat available in the molten salt is not used for heating per se, it is rather used to drive a Rankine cycle that produce electric work. A Rankine cycle is a heat engine that take high-pressure steam and expands it in a turbine almost to the dew point of the vapor. Subsequently, this saturated vapor is condensed to liquid water and then pumped into a high-pressure boiler. Efficiency of a Rankine cycle like this one is typically in the range of 40–60% relative to the Carnot efficiency, depending on irreversibilities of the system. Several mesures can be taken to make the process more efficient, but this is of less interest in this book. The point is that heat is gathered and stored with a very high efficiency and at high temperature for then to be released as electric work and some waste heat.

Example 5.2: Solar Towers and Energy Efficiency.

In this example we shall compare different solar power towers through publically available information and then evaluate the energy efficincy.

According to Wikipedia (2016), the Ivanpah solar power tower facility, California, has a maximum power of 392 MW.

Consider the heat carrying salt temperature range to be limited by 150–600°C and suppose that the steam Rankine cycle performs 50% of a Carnot cycle. The salt properties are $c_p = 1$ [kJ/kg K] and $\rho = 2.2$ [tonn/m^3].

From information of the Andasol solar power tower, there is a similar system with a peak power capacity of 50 MW, which produces 165 GWh/yr.

a) How much electric energy is produced in a 24-h cycle if all the heat is accumulated in the salt and then released?

b) What is the overall efficiency, considering that the salt reservoir is heated to 600°C before giving heat to the Carnot cycle and then cooled to 150°C while delivering all its work and then reheated (as a daily cycle)? Consider the ambient temperature to be 25°C.

c) How much salt is required (volume and mass)?

d) What is the specific energy and power of the system? Compare to Fig. 1.3.

Solutions:

a) *Considering that the two installations have the same utilization of the capacity, the energy per day is*

$$\frac{(Energy/year)_{And.}}{(P_{max})_{And.}} \frac{Yr}{day} (P_{max})_{Iva.} = \frac{165\ [\text{GWh}]}{50\ [\text{MW}]} 1/365 \left[\frac{\text{yr}}{\text{day}}\right] 392\ [\text{MW}]$$

$$= 3.54 \left[\frac{\text{MWh}}{\text{day}}\right].$$

b) *This problem requires the averaged or weighed efficiency. This is a classic unit conservation approach and is done by integrating the efficiency over the relevant temperature range and in turn by dividing this value by the temperature difference. The method is needed because if one averages the maximum and minimum efficiency, one loses information. The method stems from a column area calculation approach, where very thin columns of efficiency are calculated as trapezes; see the graphical explanation below.*

$$\bar{\varepsilon} = \frac{\int_{T_{low}}^{T_{high}} \left(1 - \frac{T_{surr}}{T}\right) dT}{\Delta_{h-m} T} = 1 - \frac{T_{surr} \ln \frac{T_{high}}{T_{low}}}{T_{high} - T_{low}} = 1 - \frac{298 \ln \frac{873}{423}}{450} = 0.52.$$

Simple average   Column average weighing   Integral average weighing

c) *The heat required from the mirrors is*

$$Q = \Delta H = \frac{W_{el}}{\varepsilon_{Carnot} \varepsilon_{Rank.}} = \frac{3.54 \text{ [MWh/d]}}{0.52 \cdot 0.5} = 13.5 \text{ [GWh]}.$$

d) *The amount of salt required is*

$$m_{salt} = \frac{Q}{\Delta T_{c_p}} = \frac{13\,500 \text{ [MWh]}}{450 \text{ [K]}} \frac{3600 \text{ [s/h]}}{1 \text{ [MWs/tonne K]}} = 108 \text{ [ktonne]}.$$

The volume becomes $V = \frac{m}{\rho} = \frac{108 \text{ [ktonne]}}{2.2 \text{ [tonne m}^{-3}]} = 49\,10^3 \text{ m}^3$, *which is equivalent to a basin of 37 m in three directions.*

<u>*Remark:*</u> *It is likely that the installation produces energy all day while on peak salt temperature, which improves the energy efficiency and lowers the amount of salt in the reservoir. This means higher energy efficiency and lower salt basin footprint. The take-home message is that these installations have higher energy efficiency than PV-systems and also have the ability to store the energy with more favorable peaks.*

As mentioned, heat is collected by a molten salt. This salt needs to be able to remain liquid at temperatures as low as possible and to remain stable at high temperatures. The chosen salts are typically based on nitrate NO_3^-, in combination with alkaline metals (K^+, Li^+, and Na^+) [9]. As illustrated in Example 5.2, it is the Carnot efficiency that primarily dictates the need for the temperature range. In Table 5.1, the boiling and melting temperatures

Table 5.1 Melting and boiling points at 1 atm for nitrate-based alkaline salts and their eutectic mixtures

Mineral	$LiNO_3$	$NaNO_3$	KNO_3	$K_{0.5}Na_{0.5}NO_3$	$K_{0.52}Li_{0.3}Na_{0.18}NO_3$
$T_m/°C$	261	307	334	221	120
$T_b/°C$	600	380	400	600	550
Source	[2]	[2]	[2]	[9]	[9]

are given for relevant salts, and corresponding eutectic compositions are given. When getting close to the melting temperature, the system is at risk to start to precipitate noneutectic mixes, which in turn can plug the system. Therefore one should be careful not to operate too close to the lower eutectic melting points. At higher temperatures, the salt can start to boil, and this is undesirable when the salt goes through some heat exchanger in the solar heating tower.

Another challenge at high temperatures is that the nitrate NO_3^- decomposes into nitrite NO_2^- and further into other compounds. Several means can be made to prevent this. More information about this, selection of the right types of materials, etc. is beyond the scope of this chapter.

PROBLEMS

Problem 5.1. Combined Heat and Power by CAES.

Compressed air energy storage (CAES) only makes sense in combination with access to surplus of heat. A large-scale solid oxide fuel (SOFC) facility has decided to keep the unit operating on a constant load and using natural gas. In doing so, the SOFC releases about half of its input chemical energy as heat. The SOFC unit requires cooling to remain at 700°C. This is done by using air that enters at 400°C (not lower to avoid thermal shock). The facility is set up so that during low electricity demand peaks it compresses air and dumps this air in an underground reservoir at 100°C. The compressed air is returned to the SOFC during high peaks in the electricity demand. The exhaust air from the SOFC is expanded adiabatically and reversibly on the way out of the SOFC. After this, the exhaust air is heat exchanged with the inlet air from 400 to 100°C and atmospheric pressure.

a) Sketch the process diagram.

b) What must the pressure in the pressurised air reservoir for the SOFC be (for the exhaust air to be expanded to 400°C)?

c) What are the work and heat of the different processes?

d) How does this contribute to the SOFC process?

Problem 5.2. Solar Tower Efficiency.

A solar tower has a heat reservoir that operates between 660 and 300°C. Consider a Rankine cycle that operates at half of the Carnot efficiency, coupled to 37°C externally.

a) What is the energy efficiency at the two temperatures?

b) The heat reservoir has its liquid replaced with antimony, which melts at 620°C so that the operation temperature is almost always 620°C. Because of a more complex situation with heating fluids, the overall efficiency of the Rankine and the system all over lowers. Evaluate the new efficiency if the system relative efficiency lowers to 40 and 45% of the Carnot efficiency. How does this affect the overall energy efficiency?

SOLUTIONS

Solution to Problem 5.1. Combined Heat and Power by CAES.

a) The process float scheme looks as follows:

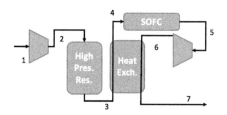

1–2: Air is compressed.

2–3: Compressed air is cooled and stored.

3–4: Compressed air is preheated in a heat exchanger.

4–5: Compressed air is heated (and oxygen partly consumed) in the SOFC.

5–6: The air expands in a turbine.

6–7: Air is cooled while preheating the SOFC inlet air stream.

b) We know the outlet pressure and the two temperatures. We can then gather the relative pressures from Appendix B, which gives them close to

$$p_5 = p_6 \frac{p_{r,5}}{p_{r,6}} = 1 \text{ [bar]} \frac{75.5}{18.4} = 4.1 \text{ [bar]}.$$

c) based on the knowledge from b), we can now know all the different states in the process scheme:

Point	$T/°C$	p/bar	p_r	$h/kJ\,kg^{-1}$
1	25	1	1	298
2	170	4.1	4.1	445
3	100	4.1	2.2	374
4	400	4.1	18.3	684
5	700	4.1	75.5	1014
6	400	1	18.4	684
7	100	1	1	374

Correspondingly we get the following energies:

$$w_{1-2} = -\Delta h_{1-2} = -147 \text{ [kJ/kg]}, \qquad q_{2-3} = \Delta h_{2-3} = -71 \text{ [kJ/kg]},$$
$$q_{3-4} = -\Delta h_{3-4} = 310 \text{ [kJ/kg]}, \qquad q_{4-5} = -\Delta h_{4-5} = 330 \text{ [kJ/kg]},$$
$$w_{5-6} = \Delta h_{5-6} = 310 \text{ [kJ/kg]}, \qquad q_{6-7} = -\Delta h_{6-7} = -310 \text{ [kJ/kg]}.$$

d) In the first step, we add 147 kJ/kg, and in the fifth step, we gain 310 kJ/kg. Thus we gain work output allover and increase the energy efficiency. The energy efficiency can be higher without the pressurized reservoir where heat is lost; however, the point here is, for instance, to run a device as a compressor at night time and as a turbine during daytime.

Solution to Problem 5.2. Solar Tower Efficiency.
a) The efficiency at the two temperatures becomes:

$$\varepsilon_{660°C} = \varepsilon_{Rank.}\varepsilon_{Carn.} = 0.5\left(1 - \frac{310 \text{ [K]}}{933 \text{ [K]}}\right) = 33.4\%,$$

$$\varepsilon_{300°C} = \varepsilon_{Rank.}\varepsilon_{Carn.}) = 0.5\left(1 - \frac{310 \text{ [K]}}{573 \text{ [K]}}\right) = 27.1\%.$$

b)

$$\varepsilon_{620°C,40\%} = \varepsilon_{Rank.}\varepsilon_{Carn.}) = 0.4\left(1 - \frac{310 \text{ [K]}}{893 \text{ [K]}}\right) = 26.1\%,$$

$$\varepsilon_{620°C,45\%} = \varepsilon_{Rank.}\varepsilon_{Carn.}) = 0.45\left(1 - \frac{310 \text{ [K]}}{893 \text{ [K]}}\right) = 29.4\%.$$

The efficiency of storing the heat in a phase change material is large and valuable, but if this in turn lowers the efficiency or conversion rate of the rest of the system (phase change is a slow process), it may not be worth the effort. This in turn means that one must look into the heat transfer of the heat transporting fluids and the belonging heat exchangers more carefully before concluding on an investment.

CHAPTER 6

Electrochemical Energy Storage

6.1 INTRODUCTION

Engineering energy storage demands basic knowledge of electrochemistry. This is as all available electric energy can be stored chemically by undergoing some sort of electrochemical reaction. Currently, the global energy consumption consists of more than 80% chemical energy (coal, oil, and gas) [10]. In part, this is because we rely heavily on chemically bound energy for transportation. In a future society, where the energy chain starts with renewable energy sources, it is likely that energy consists of as much as 80% electric energy, the rest being heat and bioenergy. Obviously, if the society is to keep up the energy usage for transportation, electrochemical energy conversion and storage are inevitable.

Energy storage is mainly about energy conversion. As conversion is a process at some rate, friction and corresponding lost work is a part of it. Accounting for efficiency is accounting for this friction. Electrochemical energy storage is accordingly all about converting electrical energy into chemical energy and vice versa. In electrochemistry, several types of friction are present. Understanding these types of friction allows us to understand energy need for entire cycles, but more importantly, it also allows understanding heat management requirements. To point it out clearly, understanding energy dissipation in electrochemistry is important because

1. It is required for appropriate dimensioning of systems to be installed.
2. It allows improvement of energy utilization.
3. The lost work is a heat source that must be accounted for in heat management and system design.

There are many different electrochemical processes where the content of this chapter comes in as crucial knowledge. Most of the industrial processes that apply electricity via an electrochemical process to make some chemical or metal product can be said to be industries of electrochemical energy storage. A great example, for instance, is aluminum. Aluminum oxide is taken into an electrolyzer to give aluminum. Aluminum can be used for many things, and under the right circumstances, it can be oxidized to give back its energy. The world's largest electrochemical process today is making sodium hydroxide by electrolysis, also known as the chlor-

Engineering Energy Storage.
DOI: 10.1016/B978-0-12-814100-7.00006-7

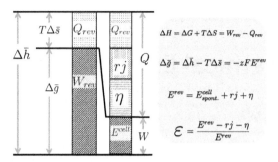

Figure 6.1 Relation between reaction energy, (enthalpy), reversible heat (product of temperature and entropy change), potential electric work, $\Delta \bar{g} = -zFE^{rev}$, the available work (cell potential) and the irreversibly lost energy for a spontaneous electrochemical cell—considering a reversible heat giving heat to the surroundings (negative heat).

alkali electrolysis. Producing hydrogen from water by electrolysis has for many decades been the most important mean for producing ultrapure hydrogen, but over the last few years, it has also become a mean to buffer large renewable energy power installations upon low electricity demand in the energy markets. Battery technology is perhaps the most commonly recognized electrochemical energy storage device seen in a daily life. Fuel cells is an example of a technology made for returning chemically stored energy back into electric energy, but still the technology appears to meet only niche markets. Electrochemical membrane processes, using ion exchange membranes, are used as purification technologies in the food industry and in desalination. It can even be reverted to give electric energy from the chemically available energy of mixing sea and river water. Electrochemical capacitors, sometimes referred to as supercapacitors, ultracapacitors, hybrid capacitors, pseudo capacitors, etc., are yet another example of how energy can be stored by electrochemical means. There are many examples, and the motivation for this chapter is to create a foundation in advanced studies of electrochemical processes; that is, when mastering this chapter, one should be able to study any industrial electrochemical process on a master level. As a bachelor within engineer, one should master all the basic terminology of electrochemical processes and be capable of taking up a job on an electrochemical industrial site.

To understand the relation between available work and the heat from an electrochemical system, Fig. 6.1 can be very useful. From a chemist point of view, all the energy converted in a system is given by the reaction enthalpy ΔH, which is a function of state. In turn, this energy is the sum of the reversible work W_{rev} and the reversible heat, Q_{rev}, which from a chemical point of view is recognized as Gibbs free energy and the product of the entropy change multiplied by the given temperature:

$$\Delta H = \Delta G + T\Delta S = W_{rev} - Q_{rev}. \tag{6.1}$$

In an electrochemical context, we are interested in the molar properties, that is, the state energies divided by the molar mass. Moreover, we are interested in the potential work or in the reversible work and its relation to reversible electrochemical cell potential, which is given by Gibbs free molar energy:

$$\Delta \bar{g} = \Delta \bar{h} - T\Delta \bar{s} = -zFE^{rev}, \tag{6.2}$$

where z is the amount of charge per reactant or, more specifically, the equivalent moles of electrons (coulombs) required per mole of reactant. For example, 2 equivalents (2 coulombs or charge of 2 moles of electrons) per mole of H_2 are exchanged when oxidizing one mole of hydrogen into two protons H^+. F is the Faraday constant.

Reversible processes obviously occur without irreversibilities. This occurs only at equilibrium, and once we move away from equilibrium, we have irreversible losses. In electrochemical systems, this means that there is a net current passing through the cell. At nonequilibrium conditions, we have irreversible processes, and the entropy production rises. From electrochemical cells there are several irreversible processes that follow from the current. Ion transport in the electrolyte due to migration of ions in the electric field is one example, expressed as ohmic losses. This is due to the friction that the ions meet when transported through the electrolyte. This is referred to as the ohmic potential drop rj after Ohm's law. In addition, the electrode processes also have irreversible energy losses, generally labeled η. This is foremost related to the friction of electrons transferred between the reactant and the electrode, later referred to as Butler–Volmer overpotential. A third friction against free movement is by diffusion of the reactant toward the surface compound, later referred to as the concentration polarization overpotential. The reversible work or the potential is related to the reversible potential and is the sum of the irreversible energy losses in the cell and the energy available for electric work outside the cell, E^{cell}:

$$E^{rev} = E^{cell}_{spont.} + rj + \eta. \tag{6.3}$$

The cell potential can be measured by a regular voltmeter in DC mode by connecting the multimeter at the poles of the cell, e.g. a battery. At open circuit, we have reversible conditions and measure the reversible potential E^{rev}. When a current is drawn, the potential lowers, and we measure the available energy as the cell potential E^{cell}. The energy efficiency ε is the ration between the two:

$$\varepsilon_{spont.} = \frac{E^{cell}_{spont.}}{E^{rev}} = \frac{E^{rev} - rj - \eta}{E^{rev}} = 1 - \frac{rj}{E^{rev}} - \frac{\eta}{E^{rev}}. \tag{6.4}$$

From Eq. (6.4) it is clear how the different irreversible energy losses lower the efficiency for a spontaneous cell and that understanding these losses is important.

When determining the heat from an electrochemical process, we can see in Fig. 6.1 that the reversible heat comes in addition to the irreversible ones. This can be both positive and negative. For instance, in Li-ion batteries, this is a cooling term during charging and a heat term during discharging.

This brief chapter of engineering elements for electrochemistry is tailored to the phenomena taking place in electrochemical energy storage devices. The explanations will be brief, but the required content for engineering is given.

$$E^{rev} = E^{cell}_{spont.} + rj + \eta, \tag{6.5}$$

$$\varepsilon_{spont.} = \frac{E^{cell}_{spont.}}{E^{rev}} = \frac{E^{rev} - rj - \eta}{E^{rev}}. \tag{6.6}$$

Electrochemistry is an order of magnitude field. Therefore it is common to apply few significant figures. In engineering electrochemistry, using logarithmic expressions is common. This indicates that many processes of electrochemical systems deal with order of magnitude. In other words, when solving problems, it is rarely necessary to apply more than two or three significant figures. It also means that many problems can easily be solved with a pen and paper or as a graphical solution, simply because of the low amount of required significant figures. Hence, we can argue that engineering electrochemistry is about making adequate approximations.

6.2 NERNST EQUATION AND THE ELECTROMOTORIC FORCE, EMF

6.2.1 The Free Energy of a Reaction

The energy of any chemical reaction is given by the enthalpy ΔH. The energy consists of two contributions, the reversible heat $T\Delta S$ and the potential energy ΔG. These are related by the equation

$$\Delta H = \Delta G + T \Delta S = W_{rev} - Q_{rev}, \tag{6.7}$$

where T is the temperature, and ΔS is the reaction entropy. Quite often and for good reasons, the equation is given in a rearranged form,

$$\Delta G = \Delta H - T \Delta S, \tag{6.8}$$

which expresses the available potential work. This term is commonly known as Gibbs free energy, free energy, chemical potential, and many similar terms. On a molar basis, giving these energies as intensive properties, we have

$$\Delta \bar{g} = \Delta \bar{h} - T \Delta \bar{s}. \tag{6.9}$$

When all compounds are at standard state, we have the standard molar reaction energy accordingly:

$$\Delta \bar{g}^o = \Delta \bar{h}^o - T \Delta \bar{s}^o. \tag{6.10}$$

The free energy expresses the maximum available work in a system, that is, the *potential energy* or potential work. The free energy can be derived from the first law of thermodynamics, which is derived in most text books on thermodynamics and in this book's Section 2.5. This potential energy is valid for any chemical reaction, here illustrated by

$$A + 2B \leftrightarrow AB_2 \tag{6.11}$$

and the molar formation (denoted f) energies per mole of the product component AB_2:

$$\Delta_f \bar{h}^o = \Delta_f \bar{h}^o_{AB_2} - \Delta_f \bar{h}^o_A - 2\Delta_f \bar{h}^o_B, \tag{6.12}$$

$$\Delta_f \bar{g}^o = \Delta_f \bar{g}^o_{AB_2} - \Delta_f \bar{g}^o_A - 2\Delta_f \bar{g}^o_B, \tag{6.13}$$

$$\Delta_f \bar{s}^o = \bar{s}^o_{AB_2} - \bar{s}^o_A - 2\bar{s}^o_B. \tag{6.14}$$

Important differences between the three properties is how they change with temperature. Whereas the enthalpic energy and entropy are close to constant with temperature. They both relate to heat capacity of the reacting components and level each other within small temperature ranges (recall that $\bar{c}_p = (\frac{\partial \bar{h}}{\partial T})_p$ and $\frac{\bar{c}_p}{T} = (\frac{\partial \bar{s}}{\partial T})_p$). The (Gibbs) free energy changes much more with temperature, as can be seen from Eqs. (6.8) and (6.9). Moreover, the ratio between the products and reactants is given by the equilibrium constant

$$K = \frac{[AB_2]}{[A][B]^2},\qquad(6.15)$$

where $[A]$ denotes the concentration of A, etc.

Example 6.1: Free Energy of a Lead Acid Accumulator.

For the reaction of a lead acid battery, the overall reaction is

$$Pb_{(s)} + PbO_{2,(s)} + 2H_2SO_{4,(aq)} \rightarrow 2PbSO_{4,(s)} + 2H_2O_{(l)}.$$

What is the standard molar free energy of formation at 298 K and 303 K, respectively? What is the standard state for the various compounds?

To solve this problem, we need to apply Eq. (6.10) and a source of information for the standard molar enthalpy and entropy of different compounds. These can be found tabulated as S.I. standard numbers [2] or equivalent:

Component	$\Delta \bar{h}_f^o / kJ\,mol^{-1}$	$\Delta \bar{g}_f^o / kJ\,mol^{-1}$	$\bar{s}^o / J\,mol^{-1}\,K^{-1}$
$Pb_{(s)}$	0	0	65
$PbO_{2,(s)}$	−277	−217	69
$H_2SO_{4,(aq\,(l))}{}^a$	−909	−744	19
$PbSO_{4,(s)}$	−920	−813	149
H_2O_l	−286	−237	70

^{a} *The values are tabulated as* $SO_{4,(aq)}^{2-}$ *since the values of the protons are 0*

	$\Delta_f \bar{h}^o$	$\Delta_f \bar{g}^o$	$\Delta_f \bar{s}^o$
Reaction	−317	−395	266

The last row in the table is calculated using Eqs. (6.12) through (6.14). The standard free energy is calculated using (6.10) at any temperature: $\Delta \bar{g}^o = \Delta \bar{h}^o - T\Delta \bar{s}^o$.

At 303 K we obtain: $\Delta \bar{g}^o = -317$ [kJ/mol] $- 303$ [K] $\cdot 0.266$ [kJ/mol K] $= -398$ [kJ/mol], *and the potential energy is slightly greater (more negative) at this temperature.*

6.2.2 The Electrochemical Free Energy

Since the free energy is a potential work, there is also a potential that must be measurable in some way. This potential is the reversible potential E^{rev}, also known as the electromotoric force (EMF). (In terms of notation, the EMF is a potential (V, J or Nm), not a force per se (N).) The reversible potential (EMF) is directly related to the free energy of the overall chemical reaction but is only considerable when the reaction is split into two parts, where each part is taken on separate electrodes. It is only in this instance that we can measure the EMF. The reversible potential is given by Nernst's equation

$$EMF = E^{rev} = -\frac{\Delta \bar{g}}{zF}, \tag{6.16}$$

where z is the amount of electrons exchanged per mole of the product component, and F is the Faraday constant. At standard state conditions, Eq. (6.16) turns into

$$EMF^o = E^o = -\frac{\Delta \bar{g}^o}{zF}. \tag{6.17}$$

Example 6.2: Free Energy of and Potential of the Lead Acid Battery.

What is the standard potential at -15, 25, 40, and 65 °C? How come this battery performs poorly when the battery gets cold?

In Example 6.1, we calculated the standard energy for the reaction of the lead acid accumulator. For this reaction, two electrons are exchanged, so that z of Eqs. (6.16) and (6.17) is 2. After calculating the free molar energy, we also obtain the cell potential.

Temperature/K	258	298	313	338
Molar free energy/kJ mol^{-1}	-386	-396	-400	-407
Potential/V	2.00	2.05	2.07	2.11

The potential increases with increasing temperature. This means that the discharge reaction is endothermic if all the potential free energy is converted to work, that is, there is only reversible heat or entropic heat.

The battery performs poorly when it gets cold, not because the potential gets lower, but because the resistance of the battery increases. This is not so transparrent in this example, however we shall look at this later in this chapter.

6.2.3 Half-Cell Reactions

In the case of electrochemical reactions, Eq. (6.11) can, be divided into to half-cell reactions. For instance, this can be:

$$Anode: \quad A \leftrightarrow A^{2+} + 2e^-; \tag{6.18}$$

$$Cathode: \quad 2B + A^{2+} + 2e^- \leftrightarrow AB_2. \tag{6.19}$$

Example 6.3: The Lead Acid Battery Half-Cell Reactions.

Having calculated the reversible and standard potential of the lead acid battery in Examples 6.1 and 6.2, we now turn the attention to half-cell standard potential. The battery discharge reaction is

$$Pb_{(s)} + PbO_{2,(s)} + 2H_2SO_{4,(aq)} \rightarrow 2PbSO_{4,(s)} + 2H_2O_{(l)}.$$

a) Determine the standard half-cell reactions for the total reaction, their reduction potentials, and the total cell potential when discharging the battery of Examples 6.1 and 6.2.

b) The half-cell potentials are given according to a reference potential. Describe this reference potential. What is the point of a reference potential?

a) *From [2] we can find half-cell reactions for each of the half-cell reactions:*

	Half-cell reduction reaction form:	E^o_{red}/V
Red.	$PbO_{2,(s)} + 4H^+ + SO^{2-}_{4,(aq)} + 2e^- \rightarrow PbSO_{4,(s)} + 2H_2O_{(l)}$	1.69
-Ox.	$PbSO_{4,(s)} + 2e^- \rightarrow Pb_{(s)} + SO^{2-}_{4,(aq)}$	−0.36
Tot.	$Pb_{(s)} + PbO_{2,(s)} + 2H_2SO_{4,(aq)} \rightarrow 2PbSO_{4,(s)} + 2H_2O_{(l)}$	2.05

This is the standard way of determining the reaction reduction potential, that is, we write both the reduction and oxidation reaction in its reduction form and then subtract the oxidation reaction (in reduction form) from the reduction reaction.

b) *All half-cell reduction reaction potentials are given with reference to the hydrogen reduction reaction:*

$$H^+ + e^- \rightarrow \frac{1}{2}H_{2,(g)}, \quad E^o = 0 \text{ V (by definition).}$$

The point of a reference potential allows us to understand how different possible competing reactions are competing with each other. Using a reference electrode, we can know the electrochemical "altitude" of a half-cell reaction, much like a barometer gives the height over the sea level for each climber on a common rope on a mountain wall. That is, the climbers know the difference in altitude between them but need a reference meter to know the exact altitude.

6.2.4 Ohm's Law: Power and Potential

Ohm's law is given by the equation

$$U = RI, \tag{6.20}$$

where U is the potential drop over a resistor with resistance R in accordance with current I. For an electrochemical cell with a current flowing through it, we can measure a cell potential E^{cell} that represents the reversible cell potential E^{rev} minus the potential lost due to the ohmic resistance in the electrolyte, $r_{cell}j$. At least for now. The measured cell potential is thus

$$E^{cell} = E^{rev} - E(j)^{ohmic} = E^{rev} - r_{cell}j, \qquad (6.21)$$

where $E(j)^{ohmic}$ is the potential loss due to current passing through a resistive electrolyte. When a galvanic cell is short circuited, the external potential drop becomes zero. From Eq. (6.21) we can now obtain the maximum current density for a spontaneous galvanic cell:

$$E^{cell} = E^{rev} - r_{cell}j_{max} = 0 \Leftrightarrow j_{max} = \frac{E^{rev}}{r_{cell}}. \qquad (6.22)$$

The potential loss is often referred to as the product of the current and the resistance as in Eq. (6.20). In electrochemistry, it is much more convenient to refer to the current density j and ohmic resistivity r, respectively. These two properties are intensive properties and usually refer to the cross-sectional area A_{cell} of a cell or electrodes. Moreover, the resistivity can be obtained from the conductivity of the electrolyte in the cell, κ_{cell}:

$$\kappa_{cell} = \frac{\delta}{r_{cell}} \Leftrightarrow r_{cell} = \frac{\delta}{\kappa_{cell}}; \qquad (6.23)$$

κ has the dimension of siemens per meter ($S\,m^{-1}$), where siemens is the inverse of ohm ($S = ohm^{-1}$). δ is the distance that the resistivity accounts for, e.g. the distance between two prallel planar plates. Moreover, the conductivity is equal to the product of the molar specific conductivity and the molar concentration:

$$\kappa = \Lambda\bar{c}, \qquad (6.24)$$

where Λ is the molar specific conductivity and $\bar{c}$ is the molar conentraiton. To a first-order approximation, the molar conductivity is proportional to the square root of the concentration and can be found tabulated,for example, in [11]. This relation is commonly known as the Debye–Hückel–Onsager relation [12], Eq. (6.25). The conductivity for a given constant concentration, one the other hand, is given by the Vogel–Tamman–Fulcher-type equation [13], Eq. (6.26).

$$\Lambda = \Lambda_0 - S\sqrt{c}, \qquad (6.25)$$

$$\kappa = \frac{A}{\sqrt{T}} \exp\left(-\frac{E_a}{R}\frac{1}{T - T_0}\right), \tag{6.26}$$

where Λ, S, A, E_a are material properties, which are unique to every system. The **essence of this paragraph** is that the conductivity of the electrolyte of electrochemical cell increases close to linearly with the concentration and close to exponentially with the temperature.

With the properties of Eqs. (6.20) through (6.23), we have all the information required to calculate the output potential and power density of a spontaneous electrochemical cell that suffer energy dissipation (lost work) from internal ohmic resistance only:

$$P^{cell} = E^{cell} j = E^{rev} j - r_{cell} j^2. \tag{6.27}$$

From Eq. (6.27) we can find the current of the maximum power of a battery. We find this when the derivative of the power with respect to current density is zero:

$$\frac{dP}{dj} = E^{rev} - 2 r_{cell} j = 0 \Leftrightarrow j_{P_{max}} = \frac{E^{rev}}{2 r_{cell}}. \tag{6.28}$$

Finally, inserting the value for the current density at maximum power, $j_{P_{max}}$, into Eq. (6.28), we obtain an expression for the power density as a function of reversible potential and cell area specific density, which can be very useful to know when evaluating cells based on open cell potential measurements:

$$P_{max} = E^{rev} j_{P_{max}} - r_{cell} j_{P_{max}}^2 = \frac{(E^{rev})^2}{2 r_{cell}} - \frac{r_{cell}(E^{rev})^2}{4 r_{cell}^2} = \frac{(E^{rev})^2}{4 r_{cell}}. \tag{6.29}$$

Example 6.4: Performance of a Lead Acid Battery.

Assuming that the specific conductivity for 1 M sulphuric acid is 30 Sm^{-1} (300 mS cm^{-1}), consider the cell of Example 6.2 with a distance between the electrodes of 5 mm.

a) Calculate the discharge cell potential and the power density for this battery for the current densities that refer to 1/4, 1/2, and 3/4 of the maximum current density.

b) For a cell that is 10 cm by 10 cm, how can this be measured experimentally? What external loads are required?

a) *To do this, we first need to determine the maximum current density. This is when the cell potential is zero ($E_{cell} = 0$). Thus from Eq. (6.21) and (6.23) we obtain*

$$j^{max} = \frac{E^{rev}}{r_{cell}} = \frac{E^{rev}\kappa}{2\delta} = \frac{2.05 \cdot 30}{0.005} = 12,300 \text{ A m}^{-2}.$$

This is the maximum current density, and the number may appear to be shockingly large (thousands of Amperes). A common notation in the electrochemical society is $A\,cm^{-2}$. Here the maximum current density is $1.23\,A\,cm^{-2}$. For aluminum electrolysis and fuel cells, the current density is in the range of 1–$5\,A\,cm^{-2}$ [14].

The required current densities for the plot are thus 0, 3,075, 6,150, 9,225, and $12,300\,A\,m^{-2}$, respectively. The values are plotted in Fig. 6.2.

$j/A\,m^{-2}$	0	3,075	6,150	9,225	12,300
E^{cell}/V	2.05	1.538	1.025	0.513	0
$p^{cell}/W\,m^{-2}$	0	4.730	6.304	4.732	0

b) First, we need to determine the internal resistance R:

$$R = r/A = \frac{\delta}{\kappa A} = \frac{0.005}{30 \cdot 0.1^2} = 0.0167 \text{ Ohm.}$$

Form Kirchof's law we know that the maximum current is when the internal and external resistances have the same values. That is, the current times the sum of the internal and external resistances is always equal to the reversible potential:

$$E^{rev} = I(R^{cell} + R^{ext.}) \Leftrightarrow R^{ext} = \frac{E^{rev}}{I} - R^{cell}.$$

Thus we can tabulate the external resistance values when the area A is $0.01\,m^2$:

I/A	0	31	62	92	123
$R^{ext.}/Ohm$	∞	0.05	0.016	0.0056	0

By changing the external resistors we can change the current and the output potential and power. From this example we can see that when the internal resistance is equal to the external, we also have the maximum power and half the cell potential. This means that half of the available electric work is lost due to the internal resistance. At maximum power, the energy efficiency is 50%. In Fig. 6.2, the cell potential and cell power density are plotted as functions of the cell current density. Here it can be seen that when the cell potential is half of the reversible or open potential, the power density peaks. This point is usually referred to as the match load power since this requires the inner and outer resistances to be equally large. With an electronic load resistance, these curves can be recorded contiuously, however knowing the basic information in this example is still very useful.

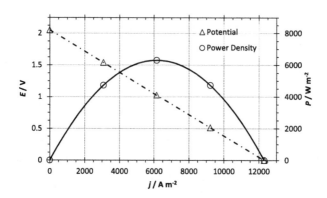

Figure 6.2 Polarization and power curves of Example 6.4.

6.3 CONCENTRATION AND NERNST EQUATION

So far, we have seen that the *standard* free energy and electrochemical potential change with temperature. Additionally, it also changes with concentration. The free energy depends on the ratio of the reactants concentration and the products concentration according to Eq. (6.30)

$$\Delta \bar{g} = \Delta \bar{g}^\circ + \bar{R}T \ln[K]. \tag{6.30}$$

Implementing reaction (6.11) into Eq. (6.30) gives

$$\Delta \bar{g} = \Delta \bar{g}^\circ + \bar{R}T \ln \left[\frac{[AB_2]}{[A][B]^2} \right]. \tag{6.31}$$

6.3.1 Activity of Components and Species

In the field of thermodynamics of aqueous systems, we need to distinguish between salts and ions. By definition, a component dissociates into several species (see, e.g., [15,16]). This is, for instance, the case for sodium chloride; the component NaCl dissociates into its ion species, Na^+ and Cl^-. Water H_2O as a component can also dissociate into species H^+ (or H_3O^+) and OH^-. For salts that dissociate entirely, that is, fluorides, chlorides (not lead and silver chloride), nitrates, sulphates (not mercury and lead sulphate), alkaline, and earth alkaline, the activity is considered to be the product of the activity of each of the species in the solution. For the salt AB_2 (similarly to reaction of Eq. (6.11), but with A being anion and B cation), the activity is given as follows:

$$a_{AB_2} = a_{A^{2-}} a_{B^-}^2. \tag{6.32}$$

Furthermore, the activity is related to the concentration. The activity of each species a_i is the product of the molar activity coefficient $\bar{\gamma}_i$ and the molar concentration $\bar{c}_i$. For ions of salt, the molar activity coefficient is defined as equal for each of the species. It is referred to as the mean molar activity coefficient $\bar{\gamma}_\pm$. Accordingly, Eq. (6.32) becomes

$$a_{AB_2} = \bar{\gamma}_\pm \bar{c}_{A^{(2-)}} \bar{\gamma}_\pm^2 \bar{c}_{B^-}^2 = \bar{\gamma}_\pm^3 \bar{c}_{A^{(2-)}} \bar{c}_{B^-}^2 = \bar{\gamma}_\pm^3 \bar{c}_{AB_2} (2\bar{c}_{AB_2})^2 = 4\bar{\gamma}_\pm^3 \bar{c}_{AB_2}^3. \quad (6.33)$$

Note that, for every AB_2, there are $2B^-$, whence the factor 2 in calculation concentrations.

The activity coefficient expresses a deviation from **ideality**. For an ideal solution, the activity of a species equals the concentration. However, particularly for concentrated solutions, the species tend to interact much more with each other than the solvent and the activity differ from the concentration. Thus this is the "social" metaphor for interpreting the activity coefficient and its property. As a thermodynamic intensive quantity, it has an energetic meaning. When speaking of the molar free energy of solution, it can be described in three levels. There are the standard state molar free energy $\bar{g}^\circ$, the molar free energy of the solution $\bar{g}$, and the auxiliary molar free energy $\bar{g}^{aux}$. The difference between the first and the second relates to the impact of change in activity, that is, $\bar{R}T\,\Delta \ln a$, whereas the difference between the second and the third relates to the deviation form ideality, that is, $\bar{R}T\,\Delta \ln[\gamma]$. Hence the activity coefficient somewhat explains a relative behavior of species and solvent on the one hand and also a well-defined intensive thermodynamic property on the other.

Moreover, when the concentration is changed, the activity coefficient changes as well. However, the concentration changes much more than the activity coefficient. As a first-order approximation, the assumption that the change in activity coefficient can be neglected compared to the change in concentration is therefore usually valid. This is illustrated in Eq. (6.34). In this example, the free energy of a component i is given. We can see that it is compared to the energy in a mixture at standard state, plus the contribution from the concentration difference, plus the contribution from the activity coefficient. If the activity coefficient hardly changes, then the ratio is close to unity, and the logarithmic term correspondingly vanishes. One can thus see that this is an order of magnitude comparison, where one primarily investigates standard state molar free energy, next, the concentration contributions, and eventually the nonideality contribution from the change in activity coefficients. Again, we see that electrochemistry is a field where calculations evaluate order of magnintude.

$$\Delta \bar{g}_i = \Delta \bar{g}_i^{\circ} + \bar{R}T \ln \frac{c_i^{\mathrm{I}}}{c_i^{\mathrm{II}}} + \bar{R}T \ln \frac{\gamma_i^{\mathrm{I}}}{\gamma_i^{\mathrm{II}}}. \tag{6.34}$$

The important point is that if we consider ideal solutions or small changes in concentrations, then the activity coefficient term is considered unity and thus negligible. Moreover, thermodynamically speaking, an ideal system is where the activity coefficients are unity rather than reversibilites.

6.3.2 EMF and Concentration

Equations (6.16), (6.30), and (6.31) together state that the open circuit potential, reversible potential, or the EMF change with concentration. Essentially, we obtain an expression for the reversible potential as a function of component and reactant concentrations:

$$E^{rev} = E^{\circ} - \frac{\bar{R}T}{zF} \ln[K]. \tag{6.35}$$

Studying constants such as the equilibrium constant, solubility constant, or simply the pH of a solution deals with comparing ratios of components. In this instance, it is most convenient to take the order of magnitude of these ratios into account, and then the Briggs logarithm is much more compatible with the decimal system than the natural logarithm. Therefore, the factor 2.303 is introduced. This is simply the number ln[10] and stems from $\ln[x] = \ln[10] \log[x]$:

$$E^{rev} = E^{\circ} - \frac{2.303 \, \bar{R}T}{zF} \log[K]. \tag{6.36}$$

Example 6.5: Concentration and the Lead Acid Battery Potential.
When discharging a lead acid battery, the concentration of sulphuric acid is lowered according to the discharged reaction

$$Pb_{(s)} + PbO_{2,(s)} + 2H_2SO_{4,(aq)} \rightarrow 2PbSO_{4,(s)} + 2H_2O_{(l)}.$$

What is the difference between the potential when fully charged and fully discharged? Consider that when fully discharged and fully charged, the density of the electrolyte is 1.06 kg/L and 1.32 kg/L, respectively. Assume an ideal solution system.

To solve this problem, we first need to define expressions for the potential when fully charged and discharged. For this, we can use either Eq. (6.35) or Eq. (6.36). With the latter, we get:

$$E_{Disch.}^{rev} = E^\circ - \frac{2.303\,\bar{R}T}{zF} \log \left[\frac{\left(a_{H_2SO_4,(aq)}^\circ\right)^2}{\left(a_{H_2SO_4,(aq)}^{Disch.}\right)^2} \right]$$

and

$$E_{Chrd.}^{rev} = E^\circ - \frac{2.303\,\bar{R}T}{zF} \log \left[\frac{a_{H_2SO_4,(aq)}^\circ}{a_{H_2SO_4,(aq)}^{Chrgd.}} \right]^2,$$

where a refers to activity at standard state ($^\circ$) when discharged (Disch.) and when charged (Chrgd.). Note that since the sulphuric acid is the only component that changes activity during the charging cycle, it also the only one that is considered in the activity contribution to the molar free energy, and hence the change in cell potential.

The difference in the potential of charged and discharged systems is then given as

$$\Delta E = E_{Chrgd.}^{rev} - E_{Disch.}^{rev} = \frac{2.303\bar{R}T}{2F} \log \left[\frac{a_{H_2SO_4}^{Chrgd.}}{a_{H_2SO_4}^{Disch.}} \right]^2$$

$$= \frac{2.303\bar{R}T}{F} \log \left[\frac{a_{H_2SO_4}^{Chrgd.}}{a_{H_2SO_4}^{Disch.}} \right].$$

For simplicity, as discussed in Section 6.3.1, we choose the specie activities to be equal to the specie concentration. Recalling Eq. (6.33) and the concentration relations for the different the species $\frac{1}{2}\bar{c}_{H^+} = \bar{c}_{SO_4^{2-}} = \bar{c}_{H_2SO_4}$, we obtain:

$$\Delta E^{Chrg/Disch} = \frac{\bar{R}T}{F} \ln \left[\frac{a_{H_2SO_4}^{Chrgd.}}{a_{H_2SO_4}^{Disch.}} \right] = \frac{\bar{R}T}{F} \ln \left[\frac{\left(\bar{c}_{H_2SO_4}^{Chrgd.}\right)^3}{\left(\bar{c}_{H_2SO_4}^{Disch.}\right)^3} \right]$$

$$= 3\frac{\bar{R}T}{F} \ln \left[\frac{\bar{c}_{H_2SO_4}^{Chrgd.}}{\bar{c}_{H_2SO_4}^{Disch.}} \right].$$

Note the switch between the natural and Briggian logarithms and recall that

$$\ln x = \ln 10 \log x = 2.303 \log x.$$

The missing piece is then determining the concentrations of sulphuric acid on a molar basis based on the mass concentrations given in the problem. For ideal solutions, we can use the expression (see Problem 6.1):

$$m_{H_2SO_4} = \frac{\rho_{H_2SO_4}\rho_{H_2O}}{\rho_{H_2O} - \rho_{H_2SO_4}} \left(\frac{m_{mix}}{\rho_{mix}} - \frac{m_{mix}}{\rho_{H_2O}} \right).$$

With the molar mass of 98 g/mol and densities of pure substances of 1 and 1.8 kg/L, we obtain concentrations of 7.3 and 1.4 mol/L. Finally,

$$\Delta E^{Chrgd./Disch.} = 3\frac{8.314 \cdot 298}{96485} \ln \left[\frac{7.3}{1.4} \right] = 0.148 \text{ V}.$$

Comment: Recall the open standard cell potential of Example 6.4 being about 2.05 V at standard conditions. We can thus see and confirm the order of importance for the two first terms of Eq. (6.34). That is, for electrochemical reactions the standard reaction formation energy is much more important than the concentration contribution, which in turn is more important than the contribution from nonideality (activity coefficients).

6.3.3 Concentration Polarization Overpotentials

Any potential that due to the reaction rate deviate from the reversible potential in addition to the ohmic potential is an overpotential η. Phenomenologically speaking, there are two types of overpotentials that are commonly considered, the concentration overpotential η_c and the reaction overpotential η_r. The latter is the subject later in Section 6.4. The overpotential represents an energy cost and is given by Eq. (6.37) for spontaneous cells and by Eq. (6.38) for nonspontaneous cells:

$$\text{Spontaneous: } \eta = -E^{rev} + E^{cell} - rj, \text{ from } E^{cell} = E^{rev} - rj - \eta \quad (6.37)$$
$$\text{Non-spontaneous: } \eta = -E^{cell} + E^{rev} - rj, \text{ from } E^{cell} = E^{rev} + rj + \eta \quad (6.38)$$

where E^{cell} is the measured cell potential, r is the ohmic resistivity, and j is the current density. When evaluating the overpotential, it is convenient to treat it in such a way that it has only one contribution. In this section, we consider only the contribution from the concentration profile near the electrode surface, that is, we neglect the reaction overpotentials η_r.

Any overpotential contributes into each of the electrodes by lowering the energy efficiency. For instance, in a spontaneous cell, the anode will have an increase in its half-cell potential and an increase in the cathode. In this way, the apparent reversible cell potential is lowered. When charging a battery or performing electrolysis, the cell potential increases beyond the sum of the ohmic potential drop and the reversible potential. Thus one must carefully find the absolute overpotential value at each electrode and add or subtract it accordingly. Hence it is convenient to express the absolute concentration polarization overpotential. This is given in the following equation, where $\bar{c}_{surf}$ and $\bar{c}_{bulk}$ are the concentrations at the electrode surface and in the bulk, respectively:

$$\eta_{\bar{c}} = \frac{\bar{R}T}{zF} \left| \ln \frac{\bar{c}_{surf}}{\bar{c}_{bulk}} \right|. \tag{6.39}$$

In Section 6.3.2, we learnt how the potential changes with concentration. This was exemplified with the concentration of sulphuric acid in a lead acid battery in Example 6.5. In this example, we saw that the total cell potential decreased by approximately 150 mV when the electrolyte concentration was lowered from 7.3 M to 1.6 M. What would then happen if the transport of sulphuric acid from the electrolyte could not immediately be transported from the electrolyte and into the surface to react? After all, a battery delivered relatively intensive currents (up to 12 kA m^{-2} according to Example 6.4). According to Fick's first law of diffusion (Eq. (6.40)), when a component is transported at rate J_i, the diffusion coefficient D_i predicts the corresponding concentration gradient $d\bar{c}/dx$:

$$J_i = -D_i \frac{d\bar{c}_i}{dx}. \tag{6.40}$$

One can argue that transport by diffusion is as much due to the concentration gradient as the concentration gradient is due to the flux of mass. Especially for nonelectrochemical systems, this is often argued. In the same manner as a heat flux is due to a temperature gradient, a mass flux by diffusion is due to the concentration gradient. However, in an electrochemical system, the mass flux is often set up by an electrochemical current, simliar to a heat flux being a result of activte heating or cooling of a surface. We can also see the similarity between the diffusion coefficient of Fick's first law and the thermal conductivity of Fourier's first law (Fourier's law for heat flux; $q = -k\frac{dT}{dx}$), which are both intrinsic properties of a system or material. The current and the mass flux relate as follows:

$$J_i = -zFj. \tag{6.41}$$

Much like in the Nernst equation, the chemical quantity and the electrochemical quantity are related by minus the Faraday constant F and the

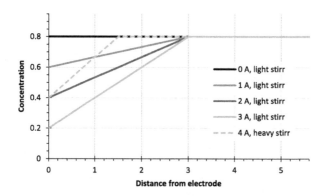

Figure 6.3 Illustrations of concentration (mol L^{-1}) profiles adjacent to the surface of an electrode consuming the reactant at different currents and electrolyte motions.

charge equivalent z. Combining Eqs. (6.40) and (6.41), we can obtain an expression for the concentration gradient as a function of the current density:

$$\frac{d\bar{c}_i}{dx} = \frac{zFj}{D_i}.$$ (6.42)

According to Eq. (6.42), the concentration next to an electrode will change linearly from the electrode surface and into the electrolyte. This is illustrated in Fig. 6.3. The solid lines show how the concentration at the surface becomes lower as the current increases while the thickness of the diffusion layer remains constant. A constant boundary thickness like that of the solid lines is typical for many systems and transport phenomena, though the boundary layer thickness is typically different for different transport phenomena. (The ratio of the boundary layer for transport of heat and mass in sea water is around 700. This is also known as the Schmidt number $Sc = \frac{v}{D}$, where v is the kinematic viscosity.) When the current increases and the boundary layer thickness remains constant, the surface concentration is lowered proportionally. This occurs to the point where the concentration of the reactant species becomes zero. At this point the current cannot become larger due to limitations in increasing the concentration differences. Regardless of how much potential is applied, the current will not increase, and this is known as the limiting current (density) j_{lim}. In this instance, the boundary thickness thus has to be lowered. This can be done simply by stirring more. In Fig. 6.3, this is shown by increasing the current by lowering the boundary layer a lot and also lowering the concentration difference.

Table 6.1 Surface concentration differences and corresponding concentration polarization differences for the currents illustrated in Fig. 6.3

I/A, Condition	0, light st.	1, light st.	2, light st.	3, light st.	4, heavy st.
$\Delta \bar{c}_s/\text{mol}\,\text{L}^{-1}$	0	0.2	0.4	0.6	0.4
$|\eta_c|/\text{mV}$	0.0	7.4	17.8	35.6	17.8

Using Eq. (6.39), we can calculate the absolute concentration overpotential $|\eta_c|$ accordingly, and this is presented in Table 6.1.

6.3.4 Liquid Junction Potential

When the two solutions I and II have the same components but different concentrations, there is a difference in chemical potential between them. This is of importance when it comes to electrochemical desalination [17], salinity gradient energy [18], and salinity gradient energy storage [19].

Consider first two salt solutions of different concentrations that are in contact with each other. This contact interface is called a **liquid junction**. The chemical potential difference, or the free energy, between two saline solutions at constant pressure is thus given by the equation

$$\Delta \bar{g} = \Delta \bar{h} - T\Delta \bar{s} = \bar{R}T \, \ln \frac{a_D}{a_C}, \tag{6.43}$$

where D is the dilute, and C is the concentrated solution. Introducing the Nernst equation, the *reversible* potential, or the electromotive force, can be expressed by the equation

$$E^{rev} = \frac{\bar{R}T}{zF} \, \ln \frac{a_C}{a_D}. \tag{6.44}$$

The activity of a salt as applied to Eq. (6.44) is traditionally expressed by the molality of the dissolved ions in relation to their activity coefficient. Moreover, the relation is given in Eq. (6.32). However, this reversible potential is not straightforwardly available. Each of the ions now can contribute with work. If equal amounts of them are transported across the interface, then there will be no net charge passing the potential free energy, and no work is actually exchanged despite the potential free energy.

We now introduce transference numbers $t_i^{\pm}$ of salt species i to our equations. In an aqueous solution, electro-neutrality is always obeyed. Electro-neutrality means that charge between species in a defined volume is neutral and that the charge transported via this volume or across a surface is balanced. That is, passing one coulomb, the sum of negative and positive charge that passes must equal 1 C. For that reason, the sum of the *transference* numbers must equal unity in a solution containing monovalent ions [8]. For instance, the consumption of one mole of Cl^- to the right in a system is compensated by the migration of one mole of Cl^- to the right or one mole of Na^+ to the left or a combination of the two alternatives such that electro-neutrality is obeyed, as described in the Eq. (6.45)

$$t_i^+ + t_i^- = 1, \tag{6.45}$$

where t_i^+ and t_i^- are the transference numbers of component i.

When considering desalination of sea water or salinity gradient energy, NaCl and water are often assumed to be the only components in the system. This can serve as a good example forward.

As mentioned, if equal amounts of charge cross the liquid junction interface, then no net work is available. It is the inequality between the two transference numbers in the interface that allows us to extract parts of the reversible potential. Such an inequality can be obtained by the use of membranes that are selective to either anions or cations. In deriving the open cell potential across a membrane or a liquid junction potential for that matter, consider therefore the cell of (6.46), where two aqueous solutions I and II are separated by a membrane j:

$$NaCl_I \ ||_j \ NaCl_{II}. \tag{6.46}$$

Across a membrane j, cationic or anionic, the measured potential is equal to the potential contribution from the anionic transport E_j^{AIT} minus the potential contribution from cationic transport E_j^{CIT}. Because there is a concentration difference between solutions I and II, we can derive Eqs. (6.47)–(6.48). These equations represent the liquid junction potential. When it comes to ion exchange membranes, these equations are named the Donnan potential or the membrane potential [4]

$$E_j^{OCP} = E_j^{AIT} - E_j^{CIT} = t_j^{Cl^-} \frac{\bar{R}T}{zF} \ln \frac{a_{Cl_I^-}}{a_{Cl_{II}^-}} - t_j^{Na^+} \frac{\bar{R}T}{zF} \ln \frac{a_{Na_I^+}}{a_{Na_{II}^+}} \tag{6.47}$$

$$= \left(t_j^{Cl^-} - t_j^{Na^+}\right) \frac{\bar{R}T}{zF} \ln \frac{a_{NaCl_I}^{1/2}}{a_{NaCl_{II}}^{1/2}} = \left(t_j^{Cl^-} - t_j^{Na^+}\right) \frac{\bar{R}T}{zF} \frac{1}{2} \ln \frac{a_{NaCl_I}}{a_{NaCl_{II}}}. \tag{6.48}$$

From electroneutrality, Eq. (6.45), we obtain the relations

$$\left(t_j^{Cl^-} - t_j^{Na^+}\right) = \left(1 - 2t_j^{Na^+}\right) = \left(2t_j^{Cl^-} - 1\right). \tag{6.49}$$

These relations can be used when combining several membranes in a stack.

6.3.4.1 Multiple Liquid Junctions and the Repeating Cell Unit

In many instances and in particular when using electrochemistry for desalination, mineral purification, and when getting energy out of water streams as in salinity gradient energy, stacking membranes are important. We now turn our attention to the thermodynamics and the electrochemistry of the *single* repeating unit cell; see Eq. (6.50) and Fig. 6.4. That is, a stack of membranes where the cationic exchange membranes (CEM) and anionic exchange membranes (AEM) are repeated in an alternating manner can be described by a repeated unit of one CEM and one AEM. Exemplifying this with sodium chloride and water, this unit cell consists, from the left, of a stream of concentrated NaCl solution, a cationic selective membrane, a stream of dilute NaCl solution, and an anionic selective membrane, respectively. The membranes here represent the liquid junctions with high selectivity of either anions or cations. As an example, the concentrated and diluted solutions here enter the cell as sea and river water, respectively. The repeatable unit cell of Fig. 6.4 is described by a classic electrochemical cell description by

$$\ldots \text{NaCl}_{(aq)}^{C} \;\|_{CM}\; \text{NaCl}_{(aq)}^{D} \;\|_{AM}\; \ldots \tag{6.50}$$

In the cell in Figs. 6.4 and 6.5, sea water is the source of the concentrated solution $\text{NaCl}_{(aq)}^{C}$, and river water is the source of the diluted solution $\text{NaCl}_{(aq)}^{D}$. Note that the anode is always considered to be on the left side in all cases regarding the RED systems described here.

For the RED unit and in the light of Eq. (6.48), when the membrane j is cationic selective membrane (CM), solution I is the concentrated, and when j is anionic selective membrane (AM), II is the concentration, respectively. We can now derive the open circuit potential of the unit cell of Fig. 6.4:

$$E_{unit}^{OCP} = \frac{1}{2}\left[\left(t_{AM}^{Na^+} - t_{AM}^{Cl^-}\right) + \left(t_{CM}^{Cl^-} - t_{CM}^{Na^+}\right)\right]\frac{\bar{R}T}{zF}\ln\frac{a_{NaCl_C}}{a_{NaCl_D}}, \tag{6.51}$$

$$E_{unit}^{OCP} = \frac{1}{2}\left(\alpha_{AM}^{Cl^-} + \alpha_{CM}^{Na^+}\right)\frac{\bar{R}T}{zF}\ln\frac{a_{NaCl_C}}{a_{NaCl_D}} = \bar{\alpha}\frac{\bar{R}T}{zF}\ln\frac{a_{NaCl_C}}{a_{NaCl_D}}, \tag{6.52}$$

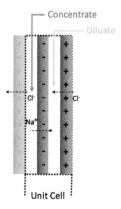

Figure 6.4 The RED unit cell consisting of (from the left) concentrated (NaCl) solution, a cation selective membrane, a diluted (NaCl) solution, and an anion selective membrane.

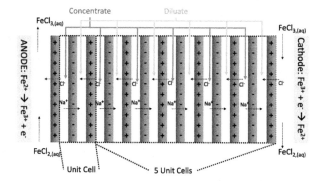

Figure 6.5 Sketch of a RED system [20]. The source of energy consists of two membranes, a brackish water solution and a concentrated saline solution forming a unit cell. These unit cells are stacked adjacent to each other to build up the cell potential. At each ends the potential built up is harvested and converted into electricity by an applied red/ox-couple, here exemplified by Fe(III)/Fe(II).

$$\alpha_{AM}^{Cl^-} = 2 \cdot t_{AM}^{Cl^-} - 1, \quad \alpha_{CM}^{Na^+} = 2 \cdot t_{CM}^{Na^+} - 1, \quad \text{and} \quad \bar{\alpha} = \left(\frac{\alpha_{AM}^{Cl^-} + \alpha_{CM}^{Na^+}}{2} \right).$$

$$(6.53)$$

By inserting Eq. (6.45) into (6.51) we obtained the expression for the open circuit potential, Eq. (6.52); $\alpha_j^{\pm}$ is the ion selectivity of the membrane j and is sometimes referred to as the perm selectivity [21] and relates to the transference number as described in Eqs. (6.53).

Note that the perm selectivity, transport number, and transference number are sometimes applied differently than here, so their original equations must be considered. For the ion selectivity of, for example, sodium in the cationic selective membrane, $\alpha_{CM}^{Na^+}$, to be more than 95 percent, the transference number must be above 0.975. A transference number of 0.9 corresponds to an ion selectivity of 80 percent, and so on. It is therefore *extremely* important not to confound these numbers. As can be seen, when the transference numbers of the two ions are equal, the available potential is zero. In the absence of selectivity, that is, when the transference numbers are equal to each other, the liquid junction potential, membrane potential, or the Donnan potential becomes nil.

In other words, lowering the transference number leads to dissipation of energy since salt is exchanged in addition to the loaded current. Therefore, increasing the selectivity of the membranes is essential in order to build up the potential of a RED stack. The transference number, and therefore the ion selectivity, is a function of the mobility, temperature, and concentration of charge carriers in the membrane. An ion selective membrane typically contains a type of ions that are locked to the membrane polymer chains, hence the signs in the membranes in the illustrations (Figs. 6.5 and 6.4). The membrane achieves its selectivity because the oppositely charged ions are mobile and able to carry the charge. Simultaneously to such loss in selectivity, the resistance of the membrane decreases. The stack open circuit potential OCP is given by the number of unit cells in the stack:

$$E_{stack}^{OCP} = N \frac{\bar{\alpha} \bar{R} T}{zF} \ln \frac{a_{NaCl_C}}{a_{NaCl_D}}, \tag{6.54}$$

where N is the number of the *pair* of membranes in the stack, and $\bar{\alpha}$ is the average membrane selectivity.

Example 6.6: The Unit Cell Performance.

For a unit cell where the membranes are 100-μm thick and separated by 200-μm thick spacers and the feed concentrations initially are 0.6 and 30 g_{NaCl}/L, use the given data to calculate:

a) The open circuit potential at first and when 5 and 10% of the potential salt transfer is transferred across the membrane. (Note that only half the salt will cross the membrane, as eventually the salt concentration is evenly distrubuted on both sides.)

b) The power density output as a function of the cross-sectional area at first and when 5 and 10% of the potential salt transfer is transferred across the membrane.

c) Calculate the fractional contribution to the unit cell internal resistance f_{R_D} from the dilute compartment accordingly.

d) In case of an experimental approach where one wants to measure the maximum power output for ten unit cells in series and with a cross-sectional area of 10 by 10 cm, what external load should be applied?

e) For processes as that exemplified here, the membrane is the major cost. Is the cross-sectional area of the membrane stack the best parameter? If not, then what area is then more relevant?

Relations between molal concentration m, mean ionic activity coefficient $\gamma_{\pm}$, and the electric conductivity κ of $NaCl_{diss}$ at 298 K.

$\dfrac{m}{\text{mol kg}^{-1}}$	0.01	0.05	0.1	0.2	0.3	0.4	0.5
$\gamma_{\pm}$	0.891[a]	0.775[a]	0.778[b]	0.735[b]	0.710[b]	0.693[b]	0.681[b,d]
$\dfrac{\kappa}{\text{S cm}^{-1}}$	0.0012[b]	0.0056[b]	0.0107[b]	0.210[c]	0.03[c]	0.04[c]	0.05[c]

[a] The ext. Debye–Hückel law [22]
[b] CRC—pp. D-169 [11]
[c] Extrapolated
[d] According to the ext. Debye–Hückel law, $\gamma_{\pm}$ is 0.471, thus underestimating $a_{NaCl_{diss}}$ by the factor 2.1

Membrane properties obtained in 0.5 m NaCl at 298 K [23].

Membrane brand	Selectivity /%	Resistance $/\Omega\,\text{cm}^2$	Thickness (wet) $/\mu\text{m}$	Conductivity $/\text{S cm}^{-1}$
CMX	92.5	3.43	181	0.0053
KESD	88.3	0.95	87	0.0092
AMX	91.0	2.65	138	0.0052
AESD	92.6	4.22	82	0.0019

To solve the first three questions (a–c), we do as follows. Here, the performances are also given for higher conversions, which can be done accordingly:

$$P_{cell\ unit} = E_{unit}^{OCP} j - \left(\frac{0.02\ \text{cm}}{0.05\ \text{S cm}^{-1}}_{(C)} + \frac{0.01}{0.0046}_{(CM)} \right.$$
$$\left. + \frac{0.02}{0.0012}_{(D)} + \frac{0.01}{0.0036}_{(AM)} \right) j^2, \tag{6.55}$$

$$P_{c.\ u.}^{0\ \%} = 0.341j - (0.40 + 2.17 + 16.67 + 2.80)j^2$$
$$= 0.341j - 22 \cdot j^2, f_{R_D} = 0.76, \tag{6.56}$$

$$P_{c.\ u.}^{5\ \%} = 0.268j - 13.4j^2, f_{R_D} = 0.60, \tag{6.57}$$

$$P_{c.\ u.}^{10\ \%} = 0.229j - 9.4j^2, f_{R_D} = 0.43, \tag{6.58}$$

$$P_{c.u.}^{20\%} = 0.139j - 7.6j^2, f_{RD} = 0.29,$$ (6.59)

$$P_{c.u.}^{40\%} = 0.091j - 6.6j^2, f_{RD} = 0.18.$$ (6.60)

c) To answer this question, we need to find the total resistance of the stack in each case of conversion. From Example 6.4 we know that the inner and outer resistances must equal each other to meet the maximum power. The specific resistance for a unit cell (as Ω cm^2) is given in Eqs. (6.55) through (6.57). The absolute resistance divided by the cross-sectional area (10 cm^2) gives the total resistances accordingly: 0.22, 0.134, and 0.094 Ω. Because there is ten in the series, each of the internal resistances need to be multiplied by 10. The external resistances should then be close to 2.2, 1.34, and 0.94 Ω accordingly. Obtainable power curves are shown in Fig. 6.6 for the different conversions. We can see that the power increases until around 10% of the salt is exchanged. In this range of conversion, the ohmic resistances lower quite rapidly in the dilute compartment. Beyond this point, the concentration difference lowers, and the drive potential lowers very rapidly. In other words, the initial part of the process is limited by the ohmic resistance in the dilute, and then the lowering in the drive potential becomes the limiting factor.

d) Because membranes are the main cost for a RED system, it is the total membrane area that is of interest rather than the cross-sectional one. For a stack of ten unit cells, we need first to multiply by the cross-sectional area and next to divide by the membrane area. Thus we obtain the following expression, where the factor two stems from two membranes (hence twice as much membranes) per unit cell:

$$P = P_{c.u.} \frac{N_{unit \, cells} A_{cross \, section} 2}{A_{cross \, section}}.$$ (6.61)

6.4 ELECTRODE REACTION KINETICS

6.4.1 The Equilibrium Reaction Rate and Constant

Away from equilibrium, products are formed. At equilibrium, products are formed as quickly as they are consumed. Thus, one way to define chemical equilibrium is when products are formed as quickly as they react back to the reactant forms. This rate is known as the equilibrium reaction rate r_0. For electrode processes, this is known as the equilibrium current I_0 or equilibrium current density j_0.

The equilibrium constant is already defined by Eq. (6.15) for reaction (6.11). Any reaction rate is given by a reaction rate constant k times the product of the concentrations of the reactants. Thus, for the formation reaction of Eq. (6.11) (the reaction forming AB_2), the reaction rate is

$$r_f = -k_f[A][B]^2.$$ (6.62)

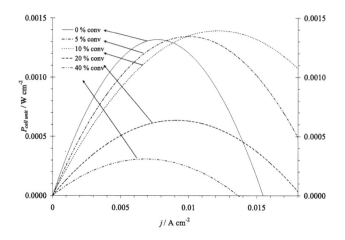

Figure 6.6 Power densities of the unit cell at 298 K as predicted in Eqs. (6.55)–(6.60).

Accordingly, the back reaction rate r_b is the consumption rate of the component AB_2 given as

$$r_b = -k_b[AB_2]. \tag{6.63}$$

Now, by definition we have equilibrium when the reaction rate is the same in both directions, that is, $r_f = r_b$:

$$k_f[A][B]^2 = k_b[AB_2], \tag{6.64}$$

$$\frac{[AB_2]}{[A][B]^2} = \frac{k_f}{k_b} = K. \tag{6.65}$$

We recognize Eq. (6.65) as an extended version of Eq. (6.15) and can learn one more thing about the equilibrium constant:

The equilibrium constant is not only the ratio between the product and the reactants; it is also the ratio between the formation and back reaction rate constants (k_f/k_b).

In terms of half-cell reactions, there is a formation reaction for the reduction reaction j_c and for the oxidation reaction j_a. According to reactions of Eqs. (6.11), (6.19), and (6.18), the Nernst relation and the formation reactions (6.18) and (6.19) can be written in a simplified way:

$$j_a = \frac{k_a[A]}{F}, \tag{6.66}$$

$$j_c = \frac{k_c[AB_2]}{2F}. \tag{6.67}$$

In general, the way to treat electrochemical calculations is to subtract the cathode reaction from the anode reaction, and hence the net current density becomes.

$$j = j_a - j_c. \tag{6.68}$$

We see that at equilibrium, $j_a = j_c$. From the equilibrium reactions of (6.18) and (6.19) we can still obtain the equilibrium constant K.

6.4.2 Butler–Volmer Overpotentials

Enough about equilibrium. When converting electric energy into chemical energy, the process operates away from equilibrium, and the net current is no longer zero. The current is then defined by Eq. (6.68). Consider the reduction reaction of Eq. (6.18) ($A^{2+} \to A$). At equilibrium when there is no net current, the component A is formed as frequent as the specie A^{2+}. However, as a current is applied to the cell in such way that a net amount of electrons are delivered to the electrode, the anodic current j_a (consuming A) becomes larger than the cathodic current j_c (producing A). This means that the anodic current is larger than the "natural" equilibrium current j_a. This increase in the current necessarily results in some sort of friction beyond the ohmic potential, rj. In other words, there is a potential change due to the friction of exchanging electrons faster than the equilibrium rate. This also applies to the reaction if electrons are removed from the electrode, and the cathodic current j_c becomes larger than the anodic current j_a. It is the reaction overpotential η_r for the electrode reaction that is evaluated in this section, whether the reaction on the electrode is of a cathodic type with a cathode overpotential η_c or vice versa, that is, an anode overpotential η_a.

To understand the nonequilibrium current, we must first introduce the equilibrium current. This is actually already done in a simplified manner in Eqs. (6.66) and (6.67). When it comes to electrode reactions, however, there is a relation between the rate coefficients k_i and the overpotential. Continuing with the reaction of Eq. (6.18), electrode equilibrium currents can be expressed by Eqs. (6.69) and (6.70) [16]. These are simplifications explained in greater detail elsewhere [16]:

$$j_a = j_0 \, \exp\left[\frac{(1-\alpha)zF}{\bar{R}T}\eta_a\right], \tag{6.69}$$

$$j_c = j_0 \, \exp\left[\frac{\alpha zF}{\bar{R}T}\eta_c\right], \tag{6.70}$$

where j_0 is the exchange current density, and α is the symmetry coefficient. The exchange current density j_0 is the equilibrium reaction rate as discussed before. It is not only different for every reaction, it is also different for every electrode material. Finding good catalysts for the electrode reactions is equivalent with finding materials with high exchange current densities, like, for example, platinum in hydrogen fuel cells. The symmetry coefficient α expresses whether the electrode reaction is favored in one direction or not. If the reaction is not favored in one direction relative to the other, then this factor is equal to 0.5, and it usually is. Introducing Eqs. (6.69) and (6.70) into Eq. (6.68), we obtain the *Butler–Volmer* equation

$$j = j_0 \left(\exp\left[\frac{(1-\alpha)zF}{\bar{R}T} \eta_a \right] - \exp\left[\frac{\alpha zF}{\bar{R}T} \eta_c \right] \right). \tag{6.71}$$

When electrons are delivered to an electrode, the anode reaction overpotential becomes negative, and the cathode positive. When electrons are removed from an electrode, the cathode reaction overpotential becomes negative, and the anode positive. A common way to deal with this is to define the reaction overpotential η_r as the cathode overpotential η_c equal to the negative anode overpotential η_a. That means that the sign convention suggests that a positive reaction overpotential refers to a oxidation reaction and a positive anode overpotential on the surface, and Eq. (6.71) becomes

$$j = j_0 \left(\exp\left[\frac{(1-\alpha)zF}{\bar{R}T} \eta_r \right] - \exp\left[-\frac{\alpha zF}{\bar{R}T} \eta_r \right] \right). \tag{6.72}$$

Looking at Eq. (6.72), we can see that there are two contributions to the reaction rate j. One is the product formation (anode or oxidation) reaction, and the other is the reactant formation (cathode or reduction) reaction. At small overpotential values η_r, the reactant formation reaction contributes significantly to limiting the overall current j, simply because the equilibrium will take some of the product and turn this into reactants again. As mentioned, this happens at small overpotential values η_r, but looking at Eq. (6.72), we can also see that once the reaction overpotential becomes sufficiently large, the product reaction becomes relatively small and can be neglected.

6.4.3 The Tafel Overpotential: An Approximation

The Butler–Volmer equation, Eq. (6.72), offers high precision at low net current densities j, but it is not very convenient to deal with when doing engineering calculations because one will have to iterate the overpotential due to the two exponential terms. Quite often in electrochemical engineering problems, like e.g. electrolysis, fuel cells, and batteries, the current density and overpotential are so large that the back reaction can be neglected. For a large positive overpotential, Eq. (6.72) can be approximated by or simplified to Eq. (6.73). This equation can be rearranged into the **Tafel overpotential equation**, Eq. (6.75):

$$j = j_0 \exp\left[\frac{(1-\alpha)zF}{\bar{R}T}\eta_r\right],\tag{6.73}$$

$$\eta_r = \frac{\bar{R}T}{(1-\alpha)zF}\ln\frac{j}{j_0},\tag{6.74}$$

$$\eta_r = -\frac{2.303\bar{R}T}{(1-\alpha)zF}\log j_0 + \frac{2.303\bar{R}T}{(1-\alpha)zF}\log j,\tag{6.75}$$

$$\eta_r = a + b\,\log j.\tag{6.76}$$

As can be seen, Eq. (6.76) is a linearized version that is much simpler to work than with Eq. (6.72). Moreover, the coefficients a and b of Eq. (6.76) are tabulated in many places, for example, Walsch [24].

Example 6.7: Reduction and Oxidation of Hydrogen on Platinum.
In this example, we compare the Butler–Volmer overpotential to the Tafel overpotential. Let us first rewrite Eq. (6.71) in the Briggian rather than in natural form. We consider α to be 0.5 and that for the hydrogen reaction, the charge equivalent per mole z is 4 (this is related to this very specific reaction kinetics and should not be used elsewhere):

$$j = j_0\left(10^{\left[\frac{(1-0.5)\cdot4\cdot96485}{2.303\cdot8.314\cdot298}\eta_a\right]} - 10^{\left[\frac{\eta_c}{0.030}\right]}\right).\tag{6.77}$$

We can now see that in this instance the Tafel slope is 30 mV (/per order of magnitude), which is in agreement with, for example, [24]. According to this reference, the exchange current density on platinum is 5 A m^{-2}.
a) Determine the current according to the given overpotentials for the Tafel behavior and the Butler–Volmer potentials.
b) When are each of the expressions most useful compared to the other?
Solutions:
a) *This is best solved by tabulating and plotting. See Fig. 6.7. From the table below it appears that the Tafel equation suggests too low overpotentials for a given current density. The error is actually the opposite. The Tafel behavior suggests too*

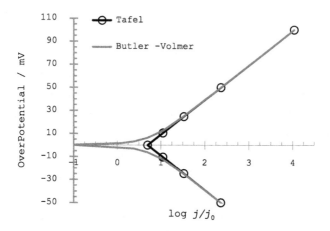

Figure 6.7 Tafel and Butler–Volmer overpotentials as functions of the logarithm of the ratio between the current density and the exchange current density according to Example 6.7.

high currents at small overpotentials due to neglecting the equilibrium current density back reaction.

j/Am^{-2}	680	58	16	7	3	2	−3	−7	−16
$\eta_{But.\,Vol.}/\text{mV}$	64	32	16	8	4	2	−4	−8	−16
η_{Tafel}/mV	64.0	31.9	14.8	3.5	–	–	–	3.5	14.8

b) For small currents, we need the Butler–Volmer relation for precision and to determine the ocerpotential at all. For higher currents, the Tafel relation offers adequate precision and is much more convenient.

6.4.4 Overpotentials for Competing Electrode Reactions

A common topic within the field of electrolysis relates to handling competing reactions. Common examples for anode and cathode reactions are chlorine/oxygen evolution on the anode and hydrogen evolution/metal deposition for the cathode. The examples are many and diverse. In this subsection the main rules for an engineers to determine the relations between the processes are laid out. In the analysis of multiple competing electrode reactions, one normally applies the Tafel overpotential reaction kinetics, Eq. (6.76).

When it comes to determining the reaction rates of two competing electrode reactions, there are two rules that apply. These rules apply both to anode reactions and cathode reactions. Also these rules are equally important.

- **The electrodes have uniform potential.** This means that the sum of the overpotential and the electrode half-cell reversible potential must be the same for the competing reactions. This potential can be measured using a reference electrode. In fact, this is the main reason *why reference electrodes are so important.* Using a reference electrode, we can determine the electrode potential. Knowing the electrode potential and the reversible half-cell electrode potential of each of the reactions allows for each of the Tafel overpotentials to be calculated, simply by subtraction. Having the Tafel kinetics parameters (coefficients *a* and *b*) for each of the reactions, we can calculate the currents and current density of each reaction. Thus we can obtain the total current to the electrode and subsequently the current efficiencies for each of the competing electrode reactions.

- **The current is the same at both electrodes.** This appears intuitive, but is still important to remember. The current distribution between several competing reactions cannot be found from this reaction rule. The only factual outcome of this rule is that once one have figured the total current on one electrode, the total current of the other electrode is the same.

Applying these rules allows for several engineering problems to be solved or exercised. One example is given in Example 6.8.

Example 6.8: The Zink Electrolysis.

Zink electrolysis is a major industrial process. In this process, zink sulphate ($ZnSO_4$) is dissolved in sulphuric acid (H_2SO_4). The overall reaction is as follows:

$$ZnSO_{4,(aq)} + H_2O_{(l)} \Rightarrow Zn_{(s)} + \frac{1}{2}O_{2,(g)} + H_2SO_{4,(aq)}.$$

The electrolyte contains 55 g_{Zn}/l and 180 $g_{H_2SO_4}$/l. We will assume ideal solutions and reaction temperature of 298 K. The applied cell potential is 3.6 V. The anode reaction overpotential η_a is 0.75 V, and the ohmic potential drop is 0.65 V. For the hydrogen evolution, the exchange current density is $3.2 \cdot 10^{-7}$ A/m^2, and the Tafel slope is 120 mV(/order of magnitude).

a) Which three electrode half-cell reactions will necessarily take place? Write these on their reduction form along with the standard reversible potential.

b) What are the anode and cathode reversible electrode potentials?

c) What are the two cathode overpotentials?

d) What are the three current densities, given that the electrodes are of a parallel plate design and a current efficiency of 95% (with respect to zinc)?

Solutions:

a) *The reactions are:*

(I) Anode–oxidation: $2H^+_{(aq)} + \dfrac{1}{2}O_{2,(g)} + 2e^- \Leftarrow H_2O_{(l)}$, $E^\circ = 1.23$ V,

(II) Cathode–main red.: $Zn^{2+}_{(aq)} + SO^{2+}_{4,(aq)} + 2e- \Rightarrow Zn_{(s)} + SO^{2+}_{4,(aq)}$,

$$E^\circ = -0.76 \text{ V},$$

(III) Cathode–side reduct.: $2H^+_{(aq)} + 2e^- \Rightarrow H_{2,(g)}$, $E^\circ = 0$ V.

b) *First, we need to determine the species concentrations. Next, we can calculate the reversible potentials. Because sulphate contributes equally on each side of the reaction arrow, the specie does not contribute to deviations from the standard potential. The calculations are all done according to Eqs. (6.32)–(6.35).*

Species	Zn^{2+}	H^+	SO^{2-}_4	Reaction	E°	E^{rev}
Standard	1	1	1	(I)	1.23	1.25
concentrations				(II)	−0.76	−0.763
Concentration	0.76	3.60	2.56	(III)	0	0.02

c) *First, we need to determine the cell potential of the cathode:*

$$E^{cath} = E^{cell} - E_{an} - rj - \eta_{an} = 3.6 - 1.25 - 0.65 - 0.75 = 0.95 \text{ V}.$$

Then we can determine the overpotential for each of the cathode reactions:

$$\eta_{H^+/H_2} = E^{cat} - E^{rev} = -0.95 - 0.02 = -0.97 \text{ V},$$
$$\eta_{Zn^{2+}/Zn} = E^{cat} - E^{rev} = -0.95 + 0.76 = -0.19 \text{ V}.$$

d) *The current I, the area A, and the current density is the same for each of the electrodes. Thus we can compare as follows:*

$$j_{anode} = j_{cathode} = j_{tot} = j_{H/H_2} + j_{Zn^{2+}/Zn} = j_{O^{2-}/O_2},$$
$$j_{tot} = f_{H/H_2}j_{tot} + f_{Zn^{2+}/Zn}j_{tot},$$
$$j_{tot} = 0.05j_{tot} + 0.95j_{tot}.$$

From this we can find the total current density from the hydrogen Tafel overpotential:

$$\eta_{H/H_2} = -b \log\left[\frac{j_{H^+/H_2}}{j_0}\right] = -b \log\left[\frac{0.05j_{tot}}{j_0}\right],$$
$$j_{tot} = \frac{1}{0.05} 10^{\left[-\frac{\eta}{b}\right]} j_0 = 20 \cdot 10^{0.97/0.120} \cdot 3.2 \cdot 10^{-7} = 775 \text{ A m}^{-2}.$$

PROBLEMS

Problem 6.1. Concentration and Density.

a) Given the single densities ρ of two pure liquid components A and B and the mass and density of the mixture ρ_{mix}, show that

$$m_A = \frac{\rho_A \rho_B}{\rho_B - \rho_A} \left(\frac{m_{mix}}{\rho_{mix}} - \frac{m_{mix}}{\rho_B} \right).$$

b) Apply this expression to fulfill the following table for B being sulphuric acid in water. The density of sulphuric acid is 1.8 kg/L, and for water it is 1 kg/L. Consider 1 L of solution.

$\rho_{mix}/\text{kg L}^{-1}$	1	1.1	1.2	1.3	1.4
m_{mix}/kg					
$m_{H_2SO_4}/\text{kg}$					
$c_{H_2SO_4}/\text{mol L}^{-1}$					

Problem 6.2. Defining Potentials.

a) How does the (Gibbs) free energy relate to the terms standard potential $E°$, the reversible potential E^{rev}, and the open circuit potential E^{OCP}?

b) Define the cell potential and the power density for a spontaneous cell as a function of cell resistance r, current density j, concentration polarization overpotential η_c, and the anode Tafel overpotential given by a, b, and $\log j$.

Problem 6.3. Tafel Kinetics from Experiments.

Someone in a laboratory measured the hydrogen evolution current on a new electrode. This is done by applying a current to the working electrode and simultaneously measuring the potential difference to a silver chloride reference potential. Then the current density on the working electrode was calculated. These values are given as follows:

Nr.	1	2	3	4	5	6	7	8	9	10
E/V_{AgCl_2}	−0.20	−0.20	−0.21	−0.22	−0.25	−0.30	−0.39	−0.44	−0.56	−0.68
$j/\text{A cm}^{-2}$	0.000	0.002	0.005	0.010	0.021	0.062	0.398	1.00	10.0	100
$\log j$										

a) Calculate the logarithmic values missing in the table.

b) Graphically, determine the Tafel kinetics: a, b, and j_0.

c) Numerically, determine the Tafel kinetics: a, b, and j_0

SOLUTIONS

Solution to Problem 6.1. Concentration and Density.

a) Given the single densities ρ of two pure liquid components A and B and the mass and density of the mixture ρ_{mix}, show that

$$m_A = \frac{\rho_A \rho_B}{\rho_B - \rho_A} \left(\frac{m_{mix}}{\rho_{mix}} - \frac{m_{mix}}{\rho_B} \right).$$

Solution. We have

$$\rho_{mix} = \frac{m_A + m_B}{V_A + V_B} = \frac{m_{mix}}{\frac{m_A}{\rho_A} + \frac{m_B}{\rho_B}} = \frac{m_{mix}}{\frac{m_A}{\rho_A} + \frac{m_{mix}}{\rho_B} - \frac{m_A}{\rho_B}}.$$

Rearranging gives

$$\frac{m_{tot}}{\rho_{mix}} = \frac{m_A}{\rho_A} + \frac{m_{tot}}{\rho_B} - \frac{m_A}{\rho_B},$$

$$\frac{m_{tot}}{\rho_{mix}} - \frac{m_{tot}}{\rho_B} = \frac{m_A}{\rho_A} - \frac{m_A}{\rho_B} = m_A \frac{\rho_B - \rho_A}{\rho_A \rho_B},$$

and finally

$$m_A = \frac{\rho_A \rho_B}{\rho_B - \rho_A} \left(\frac{m_{tot}}{\rho_{mix}} - \frac{m_{tot}}{\rho_B} \right).$$

b) Apply this expression to fulfill the following table for B being sulphuric acid in water. The density of sulphuric acid is 1.8,and for water it is 1 kg/L. Consider 1 L of solution (so that the mass is equal to the density).

$\rho_{mix}/\mathrm{kg\,L^{-1}}$	1	1.1	1.2	1.3	1.4
m_{mix}/kg	1	1.1	1.2	1.3	1.4
$m_{H_2SO_4}/\mathrm{kg}$	0	0.23	0.45	0.68	0.90
$c_{H_2SO_4}/\mathrm{mol\,L^{-1}}$	0	2.3	4.6	6.9	9.2

Solution to Problem 6.2. Defining Potentials.

a) The standard (Gibbs) free energy $\Delta \bar{g}^o$ is the negative product of the equivalent charge z, the Faraday constant, and the standard potential E^o. The contribution from mixing change both into free energy Δg and reversible potential E^{rev}. Sometimes there are other effects that make the available energy deviate even further, like the lack of supreme selectivity in the liquid junction potential. Thus the open circuit potential E^{OCP} is whatever potential we can measure from the cell.

b) The cell potential is the potential available for work during a reaction process, that is, when current is flowing. It is the work delivered to the outside of the cell, and thus we must subtract for all the internal losses:

$$E^{cell} = E^{rev} - rj - \eta_c - (a + b \cdot \log j).$$

The power density thus becomes

$$P^{cell} = E^{rev}j - rj^2 - \eta_c j - (a + b \cdot \log j)j.$$

Solution to Problem 6.3. Tafel Kinetics from Experiments.

a) This is a straightforward calculation. However, because we are going to be interested in numerical determination of Tafel kinetics in c), we add one extra line to the table:

Nr.	1	2	3	4	5	6	7	8	9	10
E/V_{AgCl_2}	−0.20	−0.20	−0.21	−0.22	−0.25	−0.30	−0.39	−0.44	−0.56	−0.68
$j/A\,cm^{-2}$	0.000	0.002	0.005	0.010	0.021	0.062	0.398	1.00	10.0	100
$\log j$	–	2.81	−2.33	−2.02	−1.67	−1.21	−0.40	0.00	1.00	2.00
$i-j$	1–2	2–3	3–4	4–5	5–6	6–7	7–8	8–9	9–10	
b_{i-j}				−0.07	−0.10	−0.12	−0.12	−0.12	−0.12	

b) Graphically, we proceed as follows, following the figures from left to right:

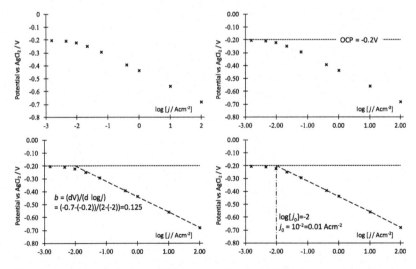

First, we plot the potential versus the logarithmic current density. Next, we determine the open cell potential. Third, we draw the Tafel slope from the region where we have a linear E-log j behavior. Here we can read the slope, the coefficient b. Finally, where the Tafel slope crosses the open cell potential line, we read (by drawing a vertical line toward the x-line) the logarithmic value of the exchange current density and in turn calculate the current density. The coefficient a is obtained from

$$a = -b\log j_o = -(-0.12) \cdot (-2) = -24.$$

c) From the tabulated data we can see that the Tafel slope is indeed 0.12 V per order of magnitude. We find this by calculating the Tafel slope between each of the measured data points;

$$b_{i-j} = \frac{\Delta_{i-j}E}{\Delta_{i-j}\log j} \underset{9-10}{\Rightarrow} \frac{-0.68 + 0.56}{2 - 1} = -0.12.$$

Rewriting the Tafel equation, we get

$$a = \eta - b\log j = (-0.68 - (-0.20)) - (-0.12 \cdot 2) = -0.48 + 24 = -0.24.$$

We can then calculate the exchange current density:

$$j_o = 10^{-\frac{a}{b}} = 10^{\frac{-0.24}{-0.12}} = 0.01 \text{ [A cm}^{-2}].$$

Remark: We can see that the graphical and numerical solutions give the same results. The numerical is faster; however, the graphical one gives a better overview and understanding of the method.

CHAPTER 7

Secondary Batteries

Batteries carry chemical energy that can be converted into electric energy whenever needed. For long time, the most commonly used batteries where disposable as primary batteries for consumer electronics. This is typically recognized as AA or AAA alkaline batteries, which can be bought in almost any store and are bought most frequently. However, rechargeable batteries seem to have become the most frequently used now. Consider your laptop, iPad (or equivalent), or your cellular phone and how much you use this compared to disposable batteries. They last for years although they are so frequently used and recharged.

Disposable batteries are called primary batteries in the *battery terminology*. Rechargeable batteries are called secondary batteries in the battery terminology. Batteries can be made from almost any type of substances as long as electrons are exchanged between a chemical substance and an electrical conductor. Because this book is about storing energy, it considers only secondary batteries. The premises for a battery is that the chemical substance must change its oxidation number and there must be two substances where one increases its oxidation number, and the other has its oxidation number decreased.

In this chapter, we introduce the most important portable secondary batteries. Different battery types are compared in terms of specific energy and power. We also explain why batteries have a limitation in range. Much of the basic electrochemistry and the lead acid battery aspects are already explained in Chapter 6, and ths treated lightly in this chapter.

At the end of this chapter, we look briefly at secondary flow batteries. This is an emerging group of batteries, where the chemical energy is stored in tanks outside the battery reactor. Because flow batteries are like rechargeable fuel cells, this also presents a good transition to the next chapter.

7.1 BATTERY TERMINOLOGY

Secondary batteries are charged and discharged. To avoid confusion, the battery technologists therefore have decided to always refer to a battery when it is discharging.

Engineering Energy Storage.
DOI: 10.1016/B978-0-12-814100-7.00007-9

A battery <u>anode</u> is the <u>oxidation</u> reaction during discharge and the reduction reaction during charging. It is termed the <u>negative</u> electrode. During discharging, the reacting element of an anode has its oxidation number increased. During charging, the reacting element has its oxidation number decreased.

A battery <u>cathode</u> is the <u>reduction</u> reaction whilst discharging and oxidation reaction during charging. It is termed the <u>positive</u> electrode. During discharge, the reacting element has its oxidation number reduced. During charging, the reacting element of the battery cathode has its oxidation number increased.

It takes current and some time to charge a battery. Those developing new materials report charging capacity as current per mass of actual electrode material. Because there are different amounts of electrode material on the anode and cathode, the denominating dimension *could* well be the current per cross-section, the current density. The denominating term for charging current is however the <u>C-rate</u>: 1C refers to a charging time of one hour from 0% state of charge (SoC) to 100% SoC, which is a full cycle. 2C refers to half an hour charging of a full cycle. Charging a battery from, for example, 30% SoC to 80% SoC (50% change in SoC) in 15 minutes is also a C-rate of 2. Thus it is leveled. Charging a full cycle in 2 hours is the C-rate C/2. Discharging gets a minus in front. The SoC refers to the maximally possible amount of charge that is shifted between the electrodes. Another term often used is the depth of discharge DoD. These terms both refer to charging window, and their sum equals unity or 100%, that is, $SoC = 100\% - DoD$.

Moreover, batteries loose capacity as they age. Therefore the term state of health (<u>SoH</u>) is introduced. SoH refers to how much charge in coulombs is available for use in the battery at a given C-rate relative to a new battery [25]. That is, an old battery is likely to have a lower SoH at a high C-rate compared to a low C-rate [26]. In particular, for Li-ion secondary batteries, some of the Li ions are fully or partially immobilized during ageing. When some Li ions are becoming very slow, they are inaccessible at higher current rates. In turn, this means that a battery will have lower SoH when considered for higher currents. This means that a battery that has been used has a lower measurable SoH at, for example, 4C compared to 1C. Thus SoH is a property that must be considered carefully for every case.

The electrodes are kept apart by a <u>separator</u>, which in many cases can be a spacer or more like a membrane. A separator is a porous physical barrier for the electrodes to come in direct contact with each other. The electrodes are typically added onto a <u>current collector</u>, which is a metal film or equivalent fit for different electrode types.

In Li-ion batteries, the anode is porous and the particles constituting the porous electrode form a special interfacial layer between the electrode particles and the electrolyte, solid–electrolyte interface (SEI). This layer is needed for the lithium ions to enter the carbon, being converted into lithium metal and then parked in the carbon structure [27]. The SEI is a layer around the anode particles that is formed during the first cycle of the battery. In addition to being needed, this layer continues to grow and will in time impede the performance of the battery [27].

The geometry of a battery van also be varied. A <u>button cell</u> is a single cell inside a disk-shaped container with a lid and is the smallest type of battery. <u>Cylindrical cells</u> consists of a very long rectangular battery that is rolled up and put into a cylindrical container with a cap. A <u>pouch cell</u> consists of several battery layers that are stacked inside a plastic bag (like vacuum packed coffee beans). <u>Prismatic cells</u> are like the pouch cells except in a rigid metal container. Some examples for Li-ion batteries are pictured in Fig. 7.6, p. 130.

Since secondary batteries can be recharged and accumulate energy, they are sometimes also referred to as electrochemical accumulators. A supercapacitor also is an electrochemical accumulator, however, not a battery.

7.2 RED-OX CELLS AND OXIDATION NUMBER

Change in oxidation number is the first key of knowledge in understanding batteries, as indicated in the former section. Therefore a set of ranking rules comes in handy. These can be best seen from the periodic table of elements. Some important facts about oxidation numbers should be established before the rules of determining the oxidation themselves. Both these facts and the ranking set of rules are given in Table 7.1. Some examples of these rules applied are given in Example 7.1.

The ranking rules for determining oxidation numbers given in Table 7.1 will give the oxidation number of most relevant elements for batteries and certainly for those treated in this book. Some elements are of course more difficult, especially when looking at some of the most modern Li-ion secondary batteries. Determining the correct oxidation number can be challenging. This is therefore treated specifically in Section 7.7.

The ultimate part of the ranking rules of oxidation number is that a rule of a lower number outranks any rule of a higher number. A classic example

Table 7.1 Oxidation number fact sheet and ranking rules

Oxidation number facts

- Every element has an oxidation number.
- For batteries, the change in oxidation number defines the electrode type.
- Oxidation numbers are written with Roman numbers after the charge sign.
- For batteries, the reaction must be separable into two half-cell reactions.
- For batteries, the half-cell reactions must exchange ions and electrons.

Oxidation number ranking rules. Lower numbered rule outrank higher numbered ones

1	The oxidation number of any pure element is always 0.
2	The oxidation number of hydrogen is always plus one $+I$.
3	The oxidation numbers of alkaline elements are always $+I$.
4	The oxidation numbers of earth alkaline elements are always $+II$.
5	The oxidation numbers of halogen elements are always $-I$.
6	The charge of the compound or ion determines the remaining oxidation number.
7	The oxidation number of oxygen is always $-II$.
8	Nondiscrete oxidation numbers are expressed with fractal Roman numbers.
9	Elements to the right and above in the periodic table of elements will win negative charge, for example, K gives electrons to Na, Na gives electrons to Al, and Cl gives electrons to F.

of this is hydrogen peroxide (H_2O_2); see Example 7.1b. In this instance, neither hydrogen nor oxygen is in their elemental form, and rule one is no longer applicable. Thus rule 2 comes into act and determines the oxidation number of hydrogen; regardless of anything else, the oxidation number for this compound must be $+I$. The next relevant rule of the ranking rules is rule 6. Since there are two hydrogen molecules, each with an oxidation number 2, the total oxidation number of oxygen is then 2, and each of the two oxygen numbers then becomes $-I$. From rule 7 we can see that oxygen usually has $-II$ as an oxidation number. Thus, when using rules lower than rule 6, oxygen always has the oxidation number of $-II$, and the other elements will deviate according to rule 6 in order to balance the charge as described by rule 6.

Example 7.1: Determination of Oxidation Numbers.

Determine the oxidation number of the elements in the following compounds:

a) H_2, O_2, and Li.

b) H_2O, H_2O_2, OH^-, and H_2O.

c) Pb, PbO_2, and $PbSO_4$.

d) $LiCoO_2$ and $Li_{1/2}CoO_2$.

Solutions:

a) *Rule 1 gives the oxidation number of all these elements: H_2^0, O_2^0, and Li^0.*

b) *Rules 2 and 6 give $H_2^{+I}O^{-II}$, which also defines rule 7. Rules 2 and 6 give $H_2^{+I}O^{-I}$. Rules 2 and 6 give $(O^{-II}H^{+I})^-$.*

c) *Rule 1 gives Pb^0. Rule 7 gives the oxidation number of oxygen, and to balance the oxidation number of lead, it must be $+IV$; $Pb^{+IV}O_2^{-II}$. Looking up for the sulphonate ion (in SI chemical data or Wikipedia) tells us that the ion charge is -2; SO_4^{2-}. Thus the oxidation number of lead needed to balance this into a neutral compound is $+2$. The oxidation number of oxygen in turn is given by rule 7. Sulphur in turn is determined by the ion charge, rule 6; $Pb^{+II}S^{+VI}O_4^{-II}$. We see that lead has oxidation numbers of 0, $+II$, and $+IV$. This is the advantage of the lead acid battery; see Section 7.6.1.*

d) *The oxidation numbers of lithium and oxygen are given as $+I$ and $-II$ by rules 3 and 7, respectively. Cobalt thus holds the remaining three positive charges by rule 6; $Li^{+I}Co^{+III}O_2^{+II}$. Likewise to the previous example, lithium and oxygen have oxidation numbers of $+I$ and $-II$, but when balancing using rules 6 and 8, cobalt now has an oxidation number of $+\frac{VII}{2}$ ($7/2 = 3.5$); $Li_{1/2}^{+I}Co^{+\frac{VII}{II}}O_{II}^{-II}$. This example illustrates how cobalt contributes in a Li-ion battery by partly changing its oxidation number.*

7.3 CHARGING AND DISCHARGE POWER AND EFFICIENCY

When discharging a battery, the available potential is the reversible or open cell potential minus the irreversible potential losses inside the battery. When charging the battery, the required voltage is the open or reversible cell potential plus the irreversible potential losses inside the battery. This is summarized in Eqs. (7.1) to (7.2). The irreversible potential losses relevant for batteries are explained in Table 7.2. The theory and nature of them are also explained and outlined in Chapter 6.

$$E_{Disch.}^{Cell} = E^{rev} - rj - \eta_{Tafel} - \eta_{Conc.}, \tag{7.1}$$

$$E_{Charging}^{Cell} = E^{rev} + rj + \eta_{Tafel} + \eta_{Conc.}. \tag{7.2}$$

The reversible potential is different from the standard potential. The difference relates to the concentration deviating from the standard concentration or standard pressure; see page 88. If intense current densities are involved in the reaction, then the reacting species will temporarily accumulate or deplete near the active surface, and it can appear as if the reversible

Table 7.2 Irreversible potential losses in batteries and their nature and behavior

$rj = \frac{\delta}{\kappa}j$	Friction of transferring charge through the battery regions, e.g. current collectors, electrode materials, electrolyte; p. 83
$\eta_{Conc.}$	Friction of component diffusion near the electrode surface; p. 91
$\eta_{Tafel} = a + b \, \log j$	Friction of electron transfer to and from reacting compound and specie; p. 103

potential is changed. The reversible potential refers to an equilibrium situation, and this concentration effect is thus regarded as an overpotential of the reaction; see page 91 and Fig. 6.3 (p. 92). When the battery is charged, the battery contains much more reactants than the product components, and therefore the reversible potential is increased at higher SoC.

The Tafel over-potential η_{Tafel} is an approximation that suits well for battery engineering problems. Its origin and deriving is explained in detail on pages 99–103, and here we only use the final equation as in Table 7.2.

The power from a discharging battery or needed to charge a battery is simply the cell potential times the current density. The efficiency is the cell potential relative to the reversible potential. Relating the energy to the potential is perhaps not so obvious, but can be shown quickly as in Eqs. (7.3) and (7.4). The overall efficiency is the product of the charging and discharging efficiencies (see Eq. (7.5)).

$$\varepsilon_{Disch.} = \frac{Energy\ out}{Energy\ stored} = \frac{P\Delta t}{P\Delta t} = \frac{E^{Cell}_{Disch.}}{E^{Rev}}, \tag{7.3}$$

$$\varepsilon_{Chrg.} = \frac{Energy\ stored}{Energy\ in} = \frac{P\Delta t}{P\Delta t} = \frac{E^{Rev}}{E^{Cell}_{Chrg.}}, \tag{7.4}$$

$$\varepsilon_{Disch./Chrg.} = \varepsilon_{Disch.}\varepsilon_{Chrg.} = \frac{E^{Cell}_{Disch.}}{E^{Rev}}\frac{E^{Rev}}{E^{Cell}_{Disch.}} = \frac{E^{Cell}_{Disch.}}{E^{Cell}_{Disch.}}. \tag{7.5}$$

7.4 BATTERY CAPACITY

The primary factors that can be said to determine a battery capacity are energy density and power density. Second, one must also account for energy efficiency, life cycle tolerance, SoC operational window, and temperature range tolerance. A brief overview of capacity factors is given in Table 7.3. These factors are indicators rather than absolute values. Also, the time frame relevant for the different technologies is indicated.

Table 7.3 Some important metric ranges for battery capacity for various well-known battery chemistries [28–30]

Capacity factor ⇓	Lead Acid	NiCd	NiMeH	ZEBRA	Li-ion
Energy/Wh kg^{-1}	20–40	40–60	50–70	100–150	150–250
Power/W kg^{-1}	5–200	10–150	10–100	150–250	100–500
Cycles/1000	1–5	1–3	1–3	1–2	1–20
Energy Eff./%	60–90	80	80	90	90–98
T-range/°C	−10–50	−20–45	−20–45	90–250	−20–50
OCV/V	2.05	1.2	1.6	2.6	3–4
SoC window/%	0–100	0–100	0–100		20–90
Relev. time frame	1940–	92-02	98-05	95-09	2005–

Among the primary capacity factors, the two first are the most obvious. For portability, as weight matters, only the volume and hence volumetric energy density are included here. When considering power versus energy, energy is usually the limiting factor, and thus a battery package usually offers sufficient power. When looking at the lower row in Table 7.3, we can see that batteries have evolved into having higher energy density rather than power. In fact, a lead acid battery offers power density similar to a modern lithium ion (Li-ion) battery, whereas the energy density has increased by one order of magnitude.

The number of charging cycles that a battery can undergo, before irreversibly loosing so much capacity that it is scrapped, is important for most commercial applications and is something one pays for. That is, cheaper batteries are often of older technology, whereas long-lasting batteries are more recently developed, and the users in need of more cycles pay for the more sophisticated and recent technology. For a system utilizing the entire capacity window on a daily basis, this might well be worth it. In other instances, it might not; for example, say, one buys a car that can drive 500 km, but on average, one never drives more than 100 km a day. This means that one uses only 0.2 weighed cycles every day. A battery offering 1500 cycles thus offers 7500 days of driving, which is equivalent to 20 years of driving. Long before 20 years, most cars have deteriorated by corrosion, and other types of wear and the more expensive batteries may not be of interest.

Energy efficiency is better in more modern batteries mainly for two reasons. The first is intrinsically related to the energy density increase. When batteries have developed from power accumulators (lead acid batteries) into energy accumulators like Li-ion batteries, one draws less fraction of the energy at the time, and the current density is lower and closer to the open cell voltage than to the voltage of the maximum power; see Fig. 6.2, p. 86. Another way of seeing this, is that a current induce some ohmic loss. The ohmic loss relative to the reversible potential defines the efficincy. When the reversible potential (related to specific energy) increases, the efficincy increases too. The other factor that improves efficiency relates to the geometric design of the batteries. Whereas the distance between the electrodes in a lead acid battery is a few millimeters, the distance between the electrodes in the more modern batteries is as small as 20 micrometers, two orders of magnitude lower. This also improves the battery efficiency.

The SoC window is advised to be limited to between 20 and 90 percent for Li-ion battery as a common practice [31]. This relates to ageing factors. Utilizing only 70% of the available energy obviously lowers the capacity of the battery. Improving this window of tolerance is thus an obvious route for capacity improvement. The capability to perform in extreme temperatures is also an indication of high capacity. Most electrochemical accumulators with organic electrolytes (Li-ion and supercapacitors) are known to reduce their expected life time by a factor of two for every increase in operation temperature by 10°C [32]. All batteries have extremely high ohmic resistance at low temperatures because of the conductivity temperature exponential relation; see (6.26), p. 84. Thus, high-temperature operation irreversibly lowers life time expectancy (number of cycles), and low-temperature operation reversibly lowers the efficiency (momentary energy and power density). Thus temperature operation range is a capacity indicator.

7.5 BATTERY FOOTPRINT

Applying a battery has especially two significant footprints that are often discussed publicly, weight and environmental emissions. In reviewing evaluations of these two topics, consider that simplifications are always made and that these must be reasonable and representative simplifications. Also, recall that, in the public debate, batteries meet conservative and biased opinions. For example, such biased oppinions can relate to why the internal combustion engine is superior to batteries, the resources, and energy required to make new batteries for the first time, why hydrogen can drive longer, etc. When following these debates, one must remember that energy storage is about the right technology, for the right purpose, at the right time, and at the right place. Looking at the Ragone chart in Fig. 1.2 (p. 9), it is pretty clear that a jet engine of a Boeing 777 is almost 100 times better in terms of energy and power density than a Li-ion battery and that therefore batteries are not fit for airliners propulsion.

7.5.1 Accumulated Weight

Most of the first electric vehicles that have met the public market are short-range cars and city buses. There are good reasons for this. Small-range cars and slow-paste buses need and carry small batteries. When a vehicle goes fast, the air friction becomes substantial, and it is not feasible for a bus to take sufficient amount of batteries along. When, for example, a car travels fast, it needs batteries to carry batteries, and the weight increases exponentially with driving range. Let us review a simple model for battery weight requirement.

For any vehicle driving about, the energy need per km can be explained by Eq. (7.6). This varies with the driving cycle [33] and several factors therein. A large heavy vehicle driving fast in the snow needs more energy to drive a kilometer than a slow driving small and light vehicle on a dry road. The energy usage or consumption is a function of the frictions (rolling and air drag) at the given driving speed plus the energy need to compensate acceleration and deceleration and auxiliary electric devices the car has in operation (air conditioning and heat pump unit is by far the most significant one).

$$Energy\ consumption = f_{Rolling} + f_{air} + E_{Accel./decel.\ losses} + E_{aux\ comp.},$$

$$\frac{dE_{drive}}{dx} = a \cdot (m_{vehicle} + m_{batt} + m_{batt\ support}) + b \cdot C_d \cdot A_{veh.} + c + d, \qquad (7.6)$$

where E_{drive} is the energy need to drive the vehicle, x is the driving distance, a is some vehicle-specific empirical coefficient, $m_{vehicle}$ is the vehicle mass with almost zero battery on board, m_{batt} is the amount of battery on board the vehicle, $m_{batt\ support}$ is the added mass to the vehicle due to the needed battery mass (e.g., structural support), b is the speed factor for the air friction depending on the drag coefficient C_d and the vehicle cross-sectional area $A_{veh.}$, c is the loss for accelerating and decelerating the vehicle, and d is the energy requirement for auxiliary components like air conditioning, lighting, electronics, etc.

In the simplification of Eq. (7.6), we neglect the energy lost by acceleration and braking deceleration c, the auxiliary components d, and also the one of added vehicle mass to support the battery weight, $m_{batt\ support}$. For

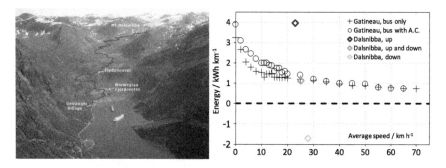

Figure 7.1 Extreme discharge and regenerating braking in Norway. The picture shows the path way from the ocean in Geiranger and 1500 meter up to Dalsnibba. The graph shows the energy per distance need for a bus with (○) and without (+) air conditioning and the uphill energy requirement (blue (dark gray in print versions) diamond), downhill regeneration (yellow (light gray in print versions) diamond), and averaged need (green (mid gray in print versions) diamond) [34].

moderate velocities, this is illustrated and validated in Fig. 7.1. Neglecting the regeneration term c is justified by comparing the distance-specific energy need of a bus driving in a practically flat urban area (near Ottawa, Canada) and probably the most extreme example Norway can offer (Geiranger–Dalsnibba) with an inclination of around 7%. This means that the regeneration is independent of velocity and terrain and can be neglected in this analysis. Neglecting the auxiliary term d is justified by reviewing the energy need for air conditioning becoming vanishingly small when driving faster than 20 km/h. The battery support term $m_{batt\ support}$ is for simplicity removed from the expression and accounted for in the battery weight term m_{batt}. Equation (7.6) is then divided by the battery specific energy e_{batt} on each side and simplifies into Eq. (7.7).

$$\frac{dE_{drive}}{dx}\frac{1}{e_{batt}} = (a \cdot (m_{vehicle} + m_{batt}) + b \cdot C_d \cdot A_{veh.})/e_{batt},$$

$$\frac{dE_{drive}}{dx}\frac{dm_{batt}}{dE_{batt}} = (a \cdot (m_{vehicle} + m_{batt}) + b \cdot C_d \cdot A_{veh.})/e_{batt},$$

$$\frac{dm_{batt}}{dx} = Am_{batt} + B, \tag{7.7}$$

$$\text{where } A = \frac{a}{e_{batt}} \text{ and } B = Am_{vehicle} + \frac{b \cdot C_d \cdot A_{veh.}}{e_{batt}}.$$

Equation (7.7) is a first-order linear equation that can be reduced to Eq. (7.8), and using zero battery mass for driving distance of zero km as

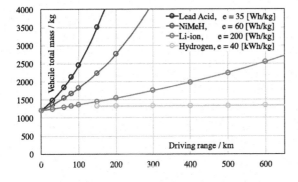

Fitting parameters
using Eq. (7.9):
$m_{vehicle} = 1200$ [kg]
$b = 0.1$ [Wh /(m^2 km kg)
$a = 0.25$ [Wh / (km kg)]
$C_d \cdot A = 0.6$ [m^2]
$m_{F.C.} = 130$ [kg]

Figure 7.2 Battery footprint in a vehicle: Weight and driving distance.

a boundary condition we obtain Eq. (7.9):

$$m_{batt}(x) = C\exp(Ax) - \frac{B}{A} = \frac{B}{A}\left(C\exp[Ax] - 1\right), \qquad (7.8)$$

$$m_{batt}(x) = \left(m_{vehicle} + \frac{b}{a} \cdot C_d \cdot A_{veh.}\right)\left(\exp\left[\frac{a}{e}x\right] - 1\right). \qquad (7.9)$$

A vehicle mass with 2–4 passengers, $m_{vehicle}$, is typically 1200–1700 kg for regular cars (up to minivans, and the air resistance constant $C_d \cdot A$ is of order 0.60–0.71 m^2 [35]. The rolling friction weight-dependent constant a can be fit and is of less importance once the driving speed becomes large. On the contrary, the energy density e_{batt} and the air resistance velocity-dependent coefficient b are of great importance for high-speed and long-range driving. The air resistance velocity function b is close to the square of the velocity. The energy density e_{batt} is included in the exponential term as a technology variable.

Fig. 7.2 compares electrochemical storage technologies to the driving range using Eq. (7.9), and the applied parameters are listed to the right of the graphical overview. Reviewing the literature, we can find several well-recognized studies that report on vehicle mass using sophisticated models for certain standardized drive cycles; see, for example, [31] and [35]. Here, the idea is to demonstrate that we can develop a simple model for a vehicle driving at a constant speed and long range. Because of braking regeneration benefits at low speed and the quadratic dependency of the coefficient b, high-speed driving is much more energy demanding. Moreover, at high range, mass accumulates exponentially, and carrying energy becomes more demanding. The problem is thus simplified into covering only the two most energy and mass demanding factors. This leaves us with a simple and valid model. In Fig. 7.2, the parameter of rolling friction weight dependency, a, and the air resistance velocity parameter were fit to Fig. 6 in [31]. With the electrical drive train included in the vehicle "energy-less" base mass, the battery mass adds exponentially from zero. Including the fuel cell in the figure required adding a fuel cell system and tank of 130 kg [36] to the base mass. From a weight point of view, we can now see that, for vehicles with a drive range of more than around 150 km, it becomes meaningless to use lead acid or NiMeH batteries. In fact, this range is close to the break even point between a fuel cell system and Li-ion batteries too. The advantage of Li-ion batteries for this range and beyond is the existing infrastructure of electric charging and the battery energy efficiency. There is a limit to how much a car with two passengers can weigh before there is no room for extra passengers and luggage. This is often argued, for certificate and license classification reasons, to be around 2200 kg. According to Fig. 7.2, a driving range of 450 km (300 miles) thus predicts the weight of the vehicle to be this weight, some 800 kg more than the fuel cell alternative. It was therefore against the reasonable weight argument that the Tesla with this driving range was introduced to the market as the 300 mile driving range car. Today, batteries outcompete fuel cells not because of weight, but because of fueling (charging) infrastructure and price. Once the hydrogen infrastructure is in place and the cost goes down, the weight arguments will favor fuel cells.

Example 7.2: How Far Can a Battery Take Itself?

Consider that a battery needs a simple vehicle and drive train of 100 kg (for now, regardless of battery weight). Use the numbers of Fig. 7.2 and Eq. (7.9) to evaluate if a battery can go forever!

Tip: Plot the mass as a function of the logarithm of the drive length.

Solution:

Equation (7.9) now becomes

$$m_{batt} = 100 \left(\exp\left[\frac{0.25}{e} x \right] - 1 \right) \text{ [kg]} \text{ and } m_{F.C.} = 230 \left(\exp\left[\frac{0.25}{e} x \right] - 1 \right) \text{ [kg]}.$$

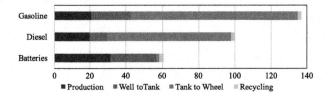

Figure 7.3 Technology relative greenhouse footprint. Life cycle greenhouse gas emissions for different propulsion technologies in a sedan considering some 200 000 km life time expectancy relative to a diesel ICE (at 100%). Values from [37].

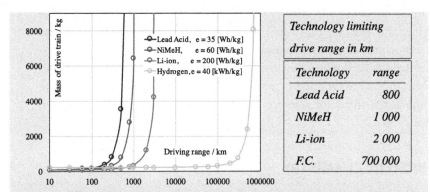

In this example we can see that any energy storage medium with conversion technology can only take itself so far before an extra tonne is nedded for the extra km. An interesting fact is that despite three times the energy density, a Li-ion battery can only go twice the distance of a NiMeH battery. This shows how non-linear (non-straightforward) these analyses are.

7.5.2 Environmental Footprint

Another common argument against new technologies relates to extraction of minerals and emission of production. Generally, it is difficult to compare technologies, and depending on how the arguments are put together, one can argue anything. However, one study done by a major car manufacturer concludes that, with the exception of acidification of the environment during mineral extraction, battery electric vehicle has a better environmental footprint than vehicles based on use of internal combustion engine (ICE) [37]. The study evaluates a typical sedan with almost 100 HP engine and a 200 000 km life range. The study looks at several different environmental aspects, but possibly the most interesting one relates to greenhouse gas footprint, which is reproduced here in Fig. 7.3. In the study, the Diesel

ICE technology serves as the references case, and the numbers included are based on a European electricity mix constituting of 45% coal power, which in 2012 was lowered to 37% and in continuous declination.

Fig. 7.3 shows that with the exception of making new batteries, BEVs have lower carbon footprint than conventional ICE vehicles. It also points at a more primary challenge; making electricity production renewable. As batteries come to take over a reasonable market share, manufacturing will be based more on recycling, and production emissions will lower. Another interesting observation that can be made is the difference between diesel Well to Tank (WtT) and gasoline WtT. Because a gasoline car has lower efficiency and requires more fuel in the Tank to Wheel (TtW) step, the WtT footprint increases also, simply as an upstream consumption propagation related to the downstream consumption. Regardless, addressing lower emissions in battery production and in the electricity supply must therefore continue as the focus for a society where greenhouse emissions are of no concern.

7.6 BATTERY CHEMISTRY

As already pointed, there are many different battery technologies. Different chemistries have different capacities. Although many of these technologies are already outdated, a brief introduction to the principle chemical reactions is convenient for engineers to be aware of.

7.6.1 Lead Acid Battery

The lead acid battery is traditionally the most commonly used battery for storing energy. It is already described extensively in Chapter 6 via the examples therein and briefly repeated here.

Lead acid battery discharge reactions		$E^0/$V
Anode	$Pb_{(s)} + H_2SO_{4,(aq)} \Rightarrow PbSO_{4,(s)} + 2H^+ + 2e^-$	1.69
Cathode	$PbO_{2,(s)} + H_2SO_{4,(aq)} + 2H^+ + 2e^- \Rightarrow PbSO_{4,(s)} + 2H_2O_{(l)}$	0.36
Total	$Pb_{(s)} + PbO_{2,(s)} + H_2SO_{4,(aq)} \Rightarrow 2PbSO_{4,(s)} + 2H_2O_{(l)}$	2.05

A lead acid battery has current collectors consisting of lead. The anode consists only of this, whereas the anode needs to have a layer of lead oxide, PbO_2. The electrodes are typically planar, and the gap between them is filled with sulphuric acid and a separator or a sulphuric acid in a gel. The separator can be a porous polymer or a woven nonconducting mesh.

One advantage of this battery is that the key component, lead (Pb), is always in a solid state. Thus the lead will only to a very low degree be lost. The disadvantage of the battery is that lead is heavy, and although the battery offers a reasonable power density, the energy density is not at all very high. Thus this battery serves as power accumulator rather than an energy accumulator. Start-up batteries are thus a good application area as one wants power rather than energy. UPS systems for electric grid support too. Stationary, that is, nonmobile, applications are another branch where the lead acid has a good market share. Lead acid is still a fairly inexpensive battery technology.

7.6.2 NiCd Batteries

The nickel cadmium, NiCd, battery was invented at the end of the 19th century. With an energy density twice of the lead acid battery, it was the first battery technology that made sense to put into portable electronics, like, for example, the first generations of laptops, drilling tools, and cellular telephones. The reactions are less straightforward than in a lead acid battery, especially, that of the cathode where the oxidation number of nickel is reduced from $+$III to $+$II.

	NiCd battery discharge reactions	E^o/V
Anode	$Cd + 2OH_{aq}^- \Rightarrow Cd(OH)_{2,s} + 2e^-$	0.4
Cathode	$2NiO(OH) + 2H_2O_l + 2e^- \Rightarrow 2Ni(OH)_{2,s} + 2OH_{aq}^-$	0.8
Total	$Cd + 2NiO(OH) + 2H_2O_l \Rightarrow 2Ni(OH)_{2,s} + Cd(OH)_{2,s}$	1.2

On the anode current collector, the cadmium is placed, and on the cathode, current collector is covered with nickel (III) oxide hydroxide. The region between the electrodes is filled with potassium hydroxide, KOH, either as a liquid or as a gel. The batteries are typically made as cylindrical cells. In the 1990s, we could buy these batteries in stores and replace them in a cell phone. Due to the relatively low energy density, keeping a spare set with you and replacing when on travel was considered an advantage.

The cathode process consists of reducing nickel. Looking at the oxidation number of nickel in the reactant form, there is two oxygen ($-$IV) and one hydrogen ($+$I) leaving nickel with an oxidation number of $+$III. In the cathode product form, nickel is associated with two hydroxide ions ($-$II), and nickel is left with an oxidation number of $+$II. Thus the oxidation number of nickel is reduced from $+$III to $+$II.

The NiCd battery is used very little today because batteries with higher energy densities are developed for a lower price. Additionally, cadmium represents hazards to biological systems, being a heavy metal, and is therefore unwanted.

7.6.3 NiMeH Batteries

An improvement of the NiCd battery is the nickel metal hydride battery. In this battery, the cadmium electrode is replaced with a hydrogen reaction, and the hydrogen is stored inside a metal electrode. The hydrogen is stored interstitially as atoms between larger atoms of a metal, thus forming a metal hydride. The metal can consist of several types of metal alloys, though it is hydrogen that undergoes the oxidation reaction during discharge. Writing MeH is an exception from the oxidation rules in that hydrogen and the metal has oxidation numbers of zero. Hydrogen, $H_{2,(g)}$, is dissolved in the solid metal and forms the metal hydride.

	NiMeH battery discharge reactions	E^o/V
Anode	$MeH_s + OH^-_{aq} \Rightarrow Me_s + H2O_l + e^-$	0.8
Cathode	$NiO(OH) + H_2O_l + e^- \Rightarrow Ni(OH)_{2,s} + OH^-_{aq}$	0.8
Total	$MeH_s + 2NiO(OH) \Rightarrow Me + 2Ni(OH)_{2,s}$	1.6

The NiMeH battery has the metal hydride attached to its anode and the nickel oxide hydride to its cathode. The electrolyte consists of potassium hydroxide, typically in an aqueous gel. The battery can be cylindrical or a prismatic one depending on what is more suitable for the application.

The NiMeH battery was widely used in portable electrodes for almost a decade, late 1990s to around 2005. The battery is perhaps best remembered for its charge *memory effect*. This effect meant that if the battery would not undergo a complete discharge, then one would loose capacity in the sense that less coulombs were available. Therefore most electronics came with charging software that would completely discharge the battery and then fully recharge while one was asleep or doing something else.

Because the NiMeH battery was so competitive in terms of energy density, it has become extremely well understood and is thus well adopted by the industry. When considering driving ranges around 10 km, a NiMeH battery competes with a Li-ion battery. This is so even up to a driving range of 50 km if considering the results in Fig. 7.2. Old technology is usually cheaper than novel ones. Sometimes more importantly, the established technology is better trusted than novel ones. When Toyota launched the Miruai, their fuel cell hybrid vehicle, the battery back remained the established NiMeH one, which is still used in the Prius, their ICE hybrid vehicle.

7.6.4 ZEBRA Batteries

The ZEBRA battery is the least commonly known battery among the batteries discussed in this book. This is probably because it only operates at high temperatures (sodium melts at 90°C), although the battery is usually operated at much higher temperatures [28,38]. This means that the battery needs to be kept warm when not in use or reheated if left unused for longer periods. ZEBRA as an acronym stands for *Zeolite Battery Research Africa*.

	ZEBRA battery discharge reaction (at 25°C)	E^0/V
Anode	$Na_l \Rightarrow Na_l^+ + e^-$	2.71
Cathode	$NiCl_{2,l} + 2Na_l^+ + 2e^- \Rightarrow Ni + 2NaCl_l$	−0.24
Total	$NiCl_{2,l} + 2Na_l \Rightarrow Ni + 2NaCl_l$	2.47

ZEBRA batteries are often made as prismatic batteries where the cathode is in the center. A carbon current collector is in a liquid cathode mixture, which is separated from the liquid anode material (Na) by a β-alumina tube. This separator conducts sodium ions and thus acts as a membrane. Higher temperatures will lower the ohmic resistance of this separator, and therefore operating temperatures above 200°C is not uncommon. The shell of the battery, some stainless steel alloy, also acts as the cathode current collector.

As the challenge of ZEBRA batteries is related to keeping the operation temperature between usage, lowering its operation temperature or at least start-up temperature is one main challenge in development. Another obvious development challenge is making a cell that tolerates thermal cycling. Whereas the latter is a material and process parameter, the former is about finding additives in the reaction chemistry and electrochemical research that harmlessly lowers the eutectic temperature.

The main market for ZEBRA batteries has been within the automotive industry where it could replace NiCd and NiMeH ones by removing environment unfriendly cadmium and increase the energy density. As can be seen in Fig. 7.4, the ZEBRA battery offers immense improvement over other battery technologies, except that Li–ion batteries have now become cheaper and significantly better. ZEBRA batteries are therefore mostly interesting for large-scale stationary energy storage, possibly in combination with distributed heating systems.

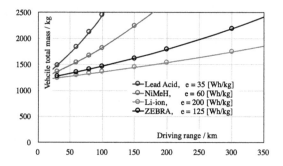

Figure 7.4 Battery vehicle weight as function of driving range, as in Fig. 7.2.

7.7 LI-ION BATTERIES

Lithium ion secondary batteries represent a wide branch of battery chemistries, which have become much more important than all other batteries and therefore need a separate section.

The lithium ion, Li-ion, battery became a commercial reality after when, in 1985, the $LiCoO_2$ (or LCO) battery was reported and patented [30]. The LCO battery thus represents the base case of the Li-ion battery family.

	LCO battery discharge reaction	E^o/V
Anode	$Li_xC \Rightarrow C + xLi_{diss.} + xe^-$	3.04
Cathode	$Li_{(1-x)}CoO_{2,s} + xLi^+_{diss.} + e^- \Rightarrow LiCoO_{2,s}$	1–1.3 V
Total	$Li_{(1-x)}CoO_{2,s} + Li_xC_s \Rightarrow LiCoO_{2,s}$	4–4.3

The standard LCO cell typically consists of an anode current collector of copper (~ 30 μm), carbon-based active anode material (~ 90 μm), a separator region (~ 20 μm), the cathode LCO active electrode material (~ 110 μm), and an aluminum current collector (~ 30 μm). The current collectors typically have the active electrode material on both sides. This is convenient when manufacturing larger cells; either one can stack layers of flat sheets so that one gets an alternating cell (Cu, An, Sep, Cath, Al, Cath, Sep, An, Cu, An, etc.), or one can take two long and narrow rectangular sheets of each type and roll up, so that when considering a radial cross section, one gets the same repeated mirror symmetric pattern. See Fig. 7.5 for inner assembly and Fig. 7.6 for final assembly pictures. A common way of assembling a Li-ion secondary battery is through a 18650-cell design, which is the standardized cell shown to the lower right in Fig. 7.6.

(A)

(B)

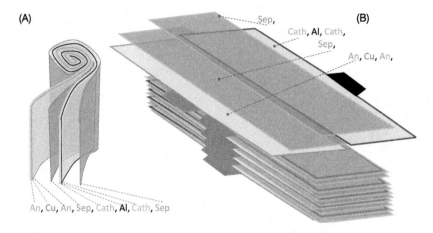

Sep,

Cath, Al, Cath,

Sep,

An, Cu, An,

An, Cu, An, Sep, Cath, Al, Cath, Sep

Figure 7.5 (A) Assembly principal for a cylindrical Li-ion battery, except the cylindrical cartridge that the battery goes inside. (B) Assembly principal for a pouch cell Li-ion battery, except the pouch that encloses the battery.

The anode material is typically based on some form of carbon particles [39]; however, several newer batteries also use silicon-based or titania-based anode material [40]. Silicon-based anodes have the advantage of taking up much less volume and weight, but expands several times throughout a cycle and are currently less matured as a technology. Titania is used for batteries that undergoes extremely fast cycles (5–15C), in part due to the capacitive properties of the material.

The separator is some porous material around 20 micrometer thick, either organic, polymeric, or like fiber glass. Several commercial batteries uses separators that irreversibly seal by melting as to prevent electric current short circuiting and further overheating, which is referred to as thermal runaways. A thermal runaway can be defined as a temperature-triggered exothermic process that produces heat faster than the battery can cool, thus leading to a temperature increase, which in turn sets yet another such reaction, eventually leading to an aggressive fire. This typically occurs first at significantly above 100°C and is described in more detail elsewhere [41]. The function of the separator when blocking the battery in reaching high temperatures is thus a security barrier inside the battery.

The cathode of Li-ion batteries consists of a very large family of transition metal oxides, using the lighter metals with partially filled d-orbitals, that is, manganese (Mn), iron (Fe), cobalt (Co), and nickel (Ni). These met-

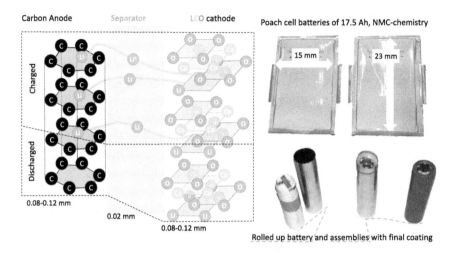

Figure 7.6 Illustrations of Li-ion battery cell description for an LCO cell (left) and commercial NMC pouch cells (upper right) and Li-ion rolled up battery assembled in a 18650-cell container (lower right). The LCO atom structure is a simplification.

als have the property of being very flexible in terms of oxidation number, and we shall evaluate this further, using cobalt as the example.

When charging a Li–ion battery, Li ions with oxidation number of +I are removed from the cathode and transported into the anode, where they are reduced into metallic ion and stored between carbon ions. The ratio of Li and carbon is one to six, and the chemistry is thus often referred to as LiC_6. This transport of lithium ions between the electrodes, enforced by an electric field, is called migration. In Fig. 7.6, this is illustrated in the upper left part. The atomic structure of the cathode consists of layers of the three constituting elements Li, Co, and O. A simplification of the atomic cathode structure is used in the example in Fig. 7.6. During charging, the Li–ions are depleted from this structure. Understanding this depletion process is one key element. Phrasing that Li is depleted and not entirely removed is essential and leads to why the cathode is charged into $Li_{(1-x)}CoO_2$. If one removes all the lithium, then the atomic crystal structure collapses and the battery can no longer be cycled. Envision the upper part of the cathode in Fig. 7.6 if more Li is removed; the structure would collapse, and Li could never react back into the structure again! Therefore, as a rule of thumb, one never removes more than half of the Li ions when charging. In fact, the x in an LCO ($\rightarrow Li_{(1-x)}CoO_2$) battery is limited to 0.55 [30].

Table 7.4 Overview of Li-ion battery cathode chemistry energy and price indication [30,42,43]

Acronym Chemistry	LCO (LiCoO$_2$)	LMO LiMn$_2$O$_4$	NMC LiNi$_\frac{1}{3}$Mn$_\frac{1}{3}$CoO$_4$	LFP LiFePO$_4$
Voltage/V	3.8–4.4	3.8–4.1	3.8–4.0	3.2–3.5
(Theor.) Sp. Energy /Wh kg^{-1}	(530) 190	(440) 150	(550) 160	(590) 160
Vol. Energy /Wh L^{-1}	560	418	260	260
Cost, 2015/16 /US\$ (NOK) Wh^{-1}	0.2 (1.5)	0.17 (1.3)	0.25 (1.9)	0.16 (1.2)
Corresp. Cycle Cost[a] /cent\$ (øre) kWh^{-1}	2–18 (13–134)	2–15 (11–114)	2–22 (17–167)	1–14 (11–107)

[a] Single cycle cost is based on 20–90 % SoC window, 80% SoH lifetime, and 2 000–20 000 cycles

Utilizing only 55% of the Li in a Li-ion battery obviously points at room for improvement. Cobalt is also an element that is expensive related to mineral reserves. In order to utilize a larger fraction of Li-ion, lower prices, improving power density and energy density has therefore led to a broad range of *cathode chemistries*. The other base chemistries and some energy metric indications are listed in Table 7.4.

When determining the cost of stored energy in a battery package, we need to know how many cycles the battery can deliver and the energy specifications of the battery package. When setting the energy specification for a battery package, we must account for SoC range used at the end of life. Li-ion batteries are often recommended to be operated at an SoC between 20 and 90% [31]. Moreover, at the end of life, the SoC is typically 80%. Beyond this point, batteries degrade very quickly, and several different safety issues arise much more frequently. Thus, when setting the energy specification needed at the end of life, we must divide the wanted energy capacity by 0.7 (accounting for SoC limitation) and then by 0.8 (accounting for end of life capacity). This means that we need to buy a battery 1.8 times as large as needed for the purpose. Once the energy specification is set, we must determine the cost of the battery package, and this relates to the amount of cycles we can have. This is always given along with the battery for different SoC windows. A modern battery package can at the very best be expected to produce up to 30 000 cycles, depending on the chemistry, temperature range, cycle frequency, SoC window used, and other specifications [30]. A tenfold in the range can be expected for different batteries, and some 2000–20 000 cycles should be a more realistic expectation depending on the given battery quality. This range in cycle toleration is the main uncertainty when determining how much we actually must pay per kWh and reflects the range in Table 7.4. Example 7.3 illustrates an example of evaluating electric energy storage for a home where the battery charge for free on the day time and saves money during peak hours.

Example 7.3: Storing Energy at Home for Power Peaks.

A household has an electricity deal that allows them to use electric energy for free (or a fixed monthly sum) as long as they use less than 1 kW at all times. For energy beyond this power (one needs energy and pays for energy rate in this instance), they must pay 15 cents per kWh. Between 5 pm and 9 pm, an average power of 1.6 kW is needed. The family considers buying a battery package to take the peak power. Under the consideration of 6 000 cycles (one cycle every evening), which battery chemistries saves money for the family? Assume prices of Table 7.4.

Solution:

First, we need to determine the energy needed beyond the free power. This is 0.6 kW (1.6–1) for 4 hours (5pm–9pm). The energy needed stored is thus 2.4 kWh.

> *Next, we compare prices:*

The cost of the extra energy is the 15 cents/kWh times the 2.4 kWh, which is 36 cents a day. This is the benchmark number that the battery technology must beat.

> *Regarding the battery package, we first look at the battery at end of life (80% SoH) and utilizing 70% of the SoC window. The battery must thus have an initial capacity of 4.29 kWh (2.4/(0.8 0.7)). This gives the price of the battery, and we can then divide by the amount of cycles (6 000) that the battery can take:*

	LCO	LMO	NMC	LFP
Battery/US$	*857*	*729*	*1071*	*686*
Energy/cent kWh^{-1}	*14.3*	*12.1*	*17.9*	*11.4*
Make or break	*Yes*	*Yes*	*No*	*Yes*

Comment: US electricity prices range 5–30 cents/kWh with place and time.

When evaluating the number of permitted cycles, care must be taken to understand the meaning of a cycle. Often a cycle refers to weighed or normalized cycles. This means that one cycle between 20 and 90% SoC refers to 0.7 normalized cycle. Thus if the battery can take, for example, 5000 normalized cycles, then the operation 20–90% SoC allows for more than 7000 cycles of that kind. Perhaps the amount of normalized cycles given for a certain SoC window is like 20–90% SoC. Moreover, if the battery is intended to age until 80% of its SoH is reached, the SoC window applied initially will be is 56% (0.7·–0.8), and the nonweighted cycle number further increases. The point is that care must be taken when choosing a battery for an application.

This book is intended to aid in understanding engineering elements rather than scientific developments in energy storage. Improvements to electrodes, separators, and electrolytes are continuously made. The description in this chapter should therefore be seen as an introduction to several scientific fields relevant for battery development. There are several things that can be done to increase capacity of Li-ion batteries and newspapers frequently report on improvements in this field of science and technology, albeit without information on potential down sides.

One case of specific energy improvement particularly worth mentioning is the Li-air battery. This battery technology is by many considered as the holy grail with secondary Li-ion battery research and consists of developing a cathode without any metal except lithium. This would increase the Li-ion battery energy density tremendously. The downside however is that it requires gas electrodes and the volume of the battery would increase ten folds, a separate air dehumidifier would be required, and currently very few cycles are reported (5–20). Having a closed container of oxygen would help on the humidity and gas feed challenges, but still the battery would be very large in terms of volume. Lowering weight, thus increasing energy density, is wanted in particular for mobile application. Decreasing the volume specific energy by a factor of ten or more (more volume same energy), the market for the battery will only be stationary energy storage where (mass-)specific energy is not very important. Therefore, increasing mass-specific energy at the cost of volumetric energy density does not make sense, at least not in the applications considered in this book. Example 7.4 evaluates the weight aspects of a Li-oxygen battery and concludes that an increase in specific energy of 50% is indeed possible by lowering the cathode specific energy by a factor of ten. This points at why many battery scientists thinks that doubling specific energy is a limitation for Li-ion batteries. In conclusion, the point here is that development in batteries must be welcomed but at the same time critically reviewed.

Example 7.4: Li-oxygen: Stop Carrying Extra Metal.

Consider an LCO battery fully charged when $x = 0.5$, that is, $Li_{1/2}CoO_2$ at SoC of 100% and $LiCoO_2$ 0 SoC. Consider that the battery-specific energy with state-of-the-art anodes and cell design is 200 Wh/kg.

A Li-air battery (Li_2O cathode) is developed, and all the Li is converted in the charging process and stored in the anode. The oxygen developed during charging is kept in a container at high pressure in relation to the battery system. (Several very clever solutions for this has been proposed. For instance, a stack of battery could be encapsulated inside a pressurized cylindrical vessel together with the oxygen and thus reduce auxiliary components.) For simplicity, assume that the cell voltage of the LCO and the proposed Li-oxygen battery is 4 V at any SoC.

a) Determine the molar energy of Li as Wh/mole.

b) What is the weight of the LCO battery per converted mole Li?

c) What is the weight of the new battery per converted mole Li? Consider only changes in cathode chemistry.

d) What is the new specific energy?

e) How many times lighter is the new cathode per mole Li?

f) How does this influence the driving range in Fig. 7.2 in Example 7.2?

Solutions:

a) *The molar energy is given by the voltage (4 V) and the Faraday constant:*

$$\bar{e}_{Li} = zFE = 1 \text{ [eq./mole}_{Li}\text{]}96485 \text{ [C/eq.]}4 \text{ [V]}/3600 \text{ [s/h]}$$
$$= 107 \text{ [Wh/mole}_{Li}\text{]}.$$

b) *The molar mass is the molar energy divided by the overall specific energy:*

$$\bar{m} = \frac{\bar{e}}{e} = \frac{107 \text{ [Wh/mol}_{Li}\text{]}}{200 \text{ [Wh/kg}_{batt.}\text{]}} = 535 \text{ [g}_{batt.}/\text{mole}_{Li}\text{]}.$$

c) *To determine this, we need to know the energy converted per mole (from a) and the mass alongside (anode, electrolyte current collectors, wrapping, etc.). In the former battery, the weight associated with a mole Li was:*

$$\bar{m}_{an+++} + 2Li + 2Co + 4O = \bar{m}_{an+++} + 14 + 118 + 64 \text{ [g}_{batt.}/\text{mole}_{Li}\text{]},$$
$$\bar{m}_{anode+++} + 196 = 535 \text{ [g}_{batt.}/\text{mole}_{Li}\text{]}.$$

Similarly to the new battery, we have:

$$m_{an.+++} + Li + 1/2O = \bar{m}_{an.+++} + 7 + 8 = m_{an.+++} + 15 \text{ [g}_{batt.}/\text{mole}_{Li}\text{]}.$$

We take an advantage of the assumption that $m_{an.+++}$ is the same for the two batteries and adjust for the difference in mass ($357 - 15 = 181$ g less per converted mole Li):

$$\bar{m}_{batt.} = 535 - 181 = 354 \text{ [g}_{batt.}/\text{mole}_{Li}\text{]}.$$

d) *The new specific energy becomes:*

$$\bar{e} = 107 \text{ [Wh/mole}_{Li}\text{]}/354 \text{ [g}_{batt.}/\text{mole}_{Li}\text{]} = 302 \text{ [Wh/kg}_{batt.}\text{]}.$$

e) *The vehicle mass lowers, and the battery needed for a driving range lowers to around the half. However, we have not considered the volume increase in this example, which would increase tremendously with including a gas. In comparison to hydrogen, we see that for long-range vehicles, hydrogen is still superior to batteries.*

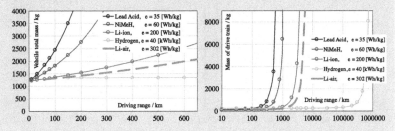

f) *From c) we see that the molar cathode weight per converted mole of Li is 196 versus 15 [g/mole$_{Li}$]. This means that the weight is lowered by a factor of more than 11. The weight of the entire battery, however, is only lowered by a factor of less than 2!*

A second case for specific energy density improvement that is widely applied is increasing the electrode thickness. This leads to lowering the mass contribution from the current collectors and the separators with electrolyte. Current collectors and electrolytes are already as thin as technically feasible. Current collectors are almost 30-μm thick, whereas a separator is a little more than 20 μm. The electrodes are both around 100 μm, and it should be clear that given this ratio, some more electrode (the useful mass) will increase the specific energy. However, adding a thicker electrode will, at any given C-rate, increase the current density, which in turn will lead to increased heating; see Eqs. (7.1) and (7.2). As mentioned, heating and correspondingly elevated temperature leads to accelerated ageing [32]. Therefore batteries intended for high power often have lower electrode thickness, which due to relatively more inactive material (current collectors and separator) in turn lower the specific energy. Example 7.5 is intended to show how lowering electrode thickness improves energy efficiency and thus lower the overall heat generation in a Li–ion battery, convenient for high C-rate cycles. A Li–ion battery with thin electrodes is often sold as a power battery and a battery with thicker electrodes as an energy battery.

Example 7.5: High Power Requires High Efficiency.
Look at Fig. 7.7 and a state of charge of 50%. Consider these properties constant in a complete charge cycle. At a charge rate of 2 C, the cross-sectional current density is 50 A m^{-2}. A battery package of 50 kWh is considered.

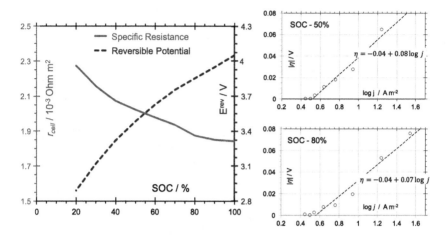

Figure 7.7 Li-ion battery cross-sectional resistivity, reversible potential, and Tafel over-potentials as functions of SoC at 298 K [39].

a) Determine the values needed for performance evaluation (Eqs. (7.1) and (7.2)).
b) What is the cell potential while charging at 2 C and 4 C?
c) What is the charging efficiency and corresponding irreversible and reversible heat? Consider charging reaction entropy of 35 J mol^{-1} K^{-1}.
d) The battery manufacturer decides to sell the batteries with electrode thickness reduced to 70% of the original one, aimed for charging at 4 C. What happens to the cell potential, efficiency, and heat production?

Solution:

a) *The specific resistance is* $2 \, m\,\Omega m^2$, *the Tafel equation* $\eta_{Tafel} = -0.05 + 0.08 \log j$, *with current density as* $A\,m^{-2}$, *and the open (reversible) cell potential is 3.6 V.*
b) *To determine the cell potential, the current density is needed at 4 C. This is twice of that at 2 C,* $100 \, A\,m^{-2}$. *The cell potentials thus becomes*

$$E_{2C}^{cell} = E^{rev} + rj + |\eta|$$
$$= 3.6 \, [V] + 0.002 \, [\Omega m^2] 50 \, [A\,m^{-2}] + |(-0.05 + 0.08 \log 50)| \, [V]$$
$$= 3.78 \, [V],$$

$$E_{4C}^{cell} = E^{rev} + rj + |\eta|$$
$$= 3.6 \, [V] + 0.002 \, [\Omega m^2] 100 \, [A\,m^{-2}] + |(-0.05 + 0.08 \log 100)| \, [V]$$
$$= 3.91 \, [V]$$

c) *The charging efficiency is the ratio between the cell potential and the reversible cell potential:* $\varepsilon = E^{cell}/E^{rev}$, *see p. 116. The heat is the power required to charge* the battery minus the power that remains in the battery:

$$\dot{Q}_{irrev.} = (E^{cell} - E^{rev})I = \left(\frac{1}{\varepsilon} - 1\right)\frac{Energy}{C - rate},$$

$$\dot{Q}_{rev.} = -\frac{T\Delta\bar{s}}{F}I = -\frac{T\Delta\bar{s}}{F \cdot E^{rev}}\frac{Energy}{\Delta t}.$$

Thus we get

Rate	$\varepsilon/\%$	$\dot{Q}_{irrev.}$/kW	$\dot{Q}_{rev.}$/kW
2 C	$\frac{3.6\,[V]}{3.78\,[V]} = 95$	$\left(\frac{1}{0.95} - 1\right)\frac{50\,[kWh]}{0.5\,[h]} = 5.3$	$\frac{-35 \cdot 298\,[J/\,mol]}{96\,485 \cdot 3.6\,[J/mol]}\frac{50\,[kWh]}{0.5\,[h]} = -3.0$
4 C	$\frac{3.6\,[V]}{3.91\,[V]} = 92$	$\left(\frac{1}{0.92} - 1\right)\frac{50\,[kWh]}{0.25\,[h]} = 17$	$\frac{-35 \cdot 298\,[J/\,mol]}{96\,485 \cdot 3.6\,[J/mol]}\frac{50\,[kWh]}{0.25\,[h]} = -6.0$

d) Changing the electrode thickness changes the current density at 4 C and the exchange current density. Both are lowered to 70%. The new current density thus becomes 70 A m^{-2}. The reduction in exchange current density changes the "a"-term in the Tafel equation, Eq. (6.76) (p. 103). The new coefficient is determined:

$$a_2 = -b \log j_{0,2} = -b \log[0.7j_{0,1}] = -b \log j_{0,1} - b \log 0.7 = a_1 - b \log 0.7.$$

With "b" equal to 0.08 V/order of magnitude, the new Tafel equation becomes

$$\eta_2 = -0.05 + 0.012 + 0.08 \log j = -0.038 + 0.08 \log j.$$

The new cell potential at 4 C thus becomes

$$E^{cell}_{4C,new} = E^{rev} + rj + |\eta|$$
$$= 3.6\,[V] + 0.002\,[\Omega m^2]70\,[A m^{-2}] + |(-0.038 + 0.08\log 70)|\,[V]$$
$$= 3.85\,[V].$$

The energy efficiency becomes

$$\varepsilon_{4C,new} = 3.6\,[V]/3.85\,[V] = 93.5\%.$$

The heat becomes

$$\dot{Q} = \left(\frac{1}{0.935} - 1\right)\frac{50\,[kWh]}{0.25\,[h]} = 14\,[kWh].$$

7.8 FLOW CELL BATTERIES

An emerging battery technology is the flow cell battery. In contrast to all the other batteries discussed in this chapter, the battery chargeable reac-

Charging   Discharging

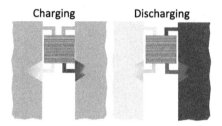

Figure 7.8 Generalized principle of a flow cell battery.

tants are kept in tanks outside of the battery reactor [44]. In this respect, a Li-ion battery can be regarded as an embedded solid rechargeable reactor. This is illustrated in Fig. 7.8, where energy difference between the flow cell solutions is indicated with darkness of blue (dark gray in print versions). The energy storage tanks are depicted on each side of the flow cell reactor. A discharged flow cell battery (left) will have the same energy level in the two containers and build up a difference gradually while charging. When discharging (right), a partial discharged solution is fed back into the storage tanks. Thus, the energy level in these tanks gradually lowers during discharge. The reactor size and design define the power (kW) capacity and the storage tanks define energy (kWh) capacity of flow cell batteries.

Flow cell batteries have the advantage of reasonably high volumetric energy density and low, if any, self-discharge rates. The volumetric energy density is in the range of 3–30 $Wh\,L^{-1}$, depending on the technology [45]. This is one order of magnitude less than Li-ion batteries; see Table 7.4. The disadvantage is low mass-specific energy content, which correspondingly would be 1–20 $Wh\,kg^{-1}$. In comparison to Li-ion batteries, this is also one order of magnitude lower. Clearly, this battery solution is not suitable for the transport sector. The power capacity can be said to still be in a developing stage, with promising outlooks [46].

When evaluating the capacity and usage of flow cell batteries, we must take into account that this is an emerging technology intended for stationary use. In this context, stationary means nonmobile. Moreover, the low volumetric energy density means that this technology is not suited for urban homes where the space of a home is an important cost factor. It is more sensible to invest into a low volume Li-ion battery package of 15–20 liters than into a flow battery system of 300+ liters, depending on the scarcity of apartment space. In this light, we can understand why flow batteries are likely to see a market in centralized energy storage systems, in less urban areas, large-scale energy storage systems, or as centralized shared energy storage systems of a neighborhood or apartment building.

There are two types of flow cell batteries. One utilizes the redox energy $\Delta \bar{g}_i$, and the other only the concentration $\Delta \bar{g}_i - \Delta \bar{g}_i^o$. This can be understood as the right side for redox flow batteries and the right side second and third terms for concentration flow batteries in Eq. (6.34), where the standard free energy term would be the redox energy and the second term the concentration energy:

$$\Delta \bar{g}_i = \Delta \bar{g}_i^o + \bar{R}T \ln \frac{c_i^I}{c_i^{II}} + \bar{R}T \ln \frac{\gamma_i^I}{\gamma_i^{II}}. \qquad (6.34)$$

(Recalling the relation between the electromotoric force (EMF) and the free energy, the potential relation should be clear:

$$EMF = E^{rev} = E^o - \frac{\bar{R}T}{zF} \ln \frac{c_i^I}{c_i^{II}} - \frac{\bar{R}T}{zF} \ln \frac{\gamma_i^I}{\gamma_i^{II}}.$$

7.8.1 RedOx Flow Batteries

Red-ox flow batteries utilize the free energy of a red-ox reaction, $\Delta \bar{g}$. It can use almost any redox reaction pair, as long as they are soluble in water. A key element however is to use transition metals that are flexible in oxidation numbers. The most common soluble metals used are vanadium, chromium, iron, cerium, and bromide. As high pH leads to precipitation of most metals (as hydroxides), the pH is usually very low, that is, less than 1. At low pH, only zink (among the nonnoble transition metals) can be reduced to a metallic form, and therefore also zink is sometimes used as a red-ox element. Depositing the metal will in time fill and block the flow path during charging. To keep it simple, zink is not considered further here.

The three elements described further, as technology examples, are vanadium, chrome, and iron [44]. For simplicity, only standard free energies E^o are considered, and it is referred to the description of concentration effects in a lead acid battery for more detailed understanding, Section 6.3 and Example 6.5, p. 88. The half-cell reduction reactions of the example elements are listed in Table 7.5.

Whereas the iron–chromium battery uses iron II and III in one chamber (reaction 1) and chromium II and III in the other reaction chamber (reaction 2), vanadium redox battery uses vanadium II and II in one chamber (reaction 3) and vanadium IV and V in the other (reaction 4). The four relevant reactions are listed in Table 7.5. The reduction reactions with a

Table 7.5 Some typically used redox single reduction reaction and corresponding reactions

	Oxidation nr.	Reduction reaction	E^o/V
1:	III→II	$Fe^{3+} + e^- \rightarrow Fe^{2+}$	0.77
2:	III→II	$Cr^{3+} + e^- \rightarrow Cr^{2+}$	−0.24
3:	III→II	$V^{3+} + e^- \rightarrow V^{2+}$	−0.26
4:	V→IV	$VO_2^+ + 2H^+ + e^- \rightarrow VO^{2+} + H_2O$	1.00

positive standard potential remain the cathode reactions, whereas the reduction reactions with a negative standard potential become the anode reactions and thus change direction and sign. The solutions are typically dissolved in sulphuric acid and separated by a proton conductive polymer membrane, which means that the ionic charge transferred between the reaction chambers are protons. For the iron chromium battery, it is important to find a membrane that rejects the iron and chromium ions. This is also so for the vanadium battery, however less important as it is vanadium of some kind in all compartments. This also point at one ageing mechanism for redox batteries, unwanted exchange of ions.

The iron chromium and vanadium flow battery total reaction becomes Eqs. (7.10) and (7.11), respectively. The part about the balancing with the sulphate ion is perhaps not so obvious, but it is based on the same principal as the lead acid battery.

$$Fe_2(SO_4)_3 + CrSO_4 \Rightarrow FeSO_4 + Cr_2(SO_4)_3, \quad E^o = 1.01 \text{ V}, \qquad (7.10)$$

$$(VO_2)_2SO_4 + H_2SO_4 + 2VSO_4 \Rightarrow 2VO(SO_4) + V_2(SO_4)_3 + H_2O,$$
$$E^o = 1.26 \text{ V}. \qquad (7.11)$$

We see that the vanadium battery has a higher standard cell potential, however, smaller than that in a lead acid or a Li-ion battery. The advantage of the redox flow batteries is that they are scalable and suitable for remote storage. The main loss in energy is the ohmic resistance and the resistance of pumping the fluids through the reacting chambers, which are very thin to lower the ohmic resistance. When the flow compartment thickness lowers, the ohmic resistance increases proportionally, whereas the pumping pressure drop increases inversely to the thickness by a power of three. There is an optimum between these losses, and as ohmic resistance and pumping pressure both lower with increasing temperature, it is clear that increased temperature is benign to the point where the membrane disintegrates. In fact, the best performing reactors are based on controlled laminar flow, where the membrane is omitted, and the anode and cathode fluid flows adjacent through the reactor [46]. Such reactors are only possible for the vanadium battery since vanadium cross-over is less important than iron or chromium cross-over.

7.8.2 Concentration Flow Batteries

Concentration cells represent flow batteries. Electrodialysis (ED) can be used to build up concentration differences between two solutions. When operating the same cell in reverse mode, reverse electrodialysis (RED), the energy can be returned. This technology is explained in Section 6.3.4, p. 93.

Whereas the energy density of redox flow cell batteries is 20–30 Wh/L, the energy density of concentration cells using RED is 2–5 Wh/L. Yet, the technology has the advantage of cheap and environmental friendly energy storage solutions (NaCl and water). Considering two very large basins of salty water and a suitably sized ED/RED reactor, energy and power can be provided for. Suitable market is large-scale grid energy storage in flat areas where pumped hydro is not an option. A salinity differences of 60 g/L have an osmotic pressure close to a water column of 600 m. This illustrates how potent this technology is for large-scale grid energy storage.

PROBLEMS

Problem 7.1. Battery Characteristics.
A battery has an SoC of 20%. The battery is subject to charging up to an SoC of 80% in 40 minutes. At 1 C, the current density across the electrolyte is 20 A m^{-2}. Cell E-r characteristics can be found in Fig. 7.7. Consider Tafel kinetics of $\eta = -0.02 + 0.03 \log j$ with current density units of A m^{-2}.
a) What are the C-rate and current density?
b) What are the open cell potential and specific ohmic resistance at open circuit potential when starting the charging?
c) Determine the cell potential at the start of the charging cycle.

Problem 7.2. Long-Range Battery Vehicles.
Several battery enthusiasts argue and point at the possibility to fast charge battery cars along the road. In practice, a supercharger can deliver some 500 MW electricity and charge a vehicle in 10 minutes. A car driving on average 90 km/h need a power of at least 20 kW. Consider a car doing this

for 5 hours. Let us assume that the E-r values at 50% are representative as a first-order approximation in this instance (open cell potential of 3.65 V and specific resistance of 2 m Ωm^2). Consider Tafel kinetics of $\eta = -0.02 + 0.03 \log j$, with current density units of A m^{-2}. At 1 C, the current density across the electrolyte is 20 A m^{-2}

a) Determine the C-rate for a full cycle.
b) Determine the current density.
c) Determine the cell potential.
d) Determine the efficiency.
f) Determine the required electric power.
g) Determine the surplus irreversible heat production (kW).

Problem 7.3. High–Power Battery Design.

A Li-ion battery has an open cell potential of 3.6 and is being charged at 2 C (equivalent to 50 A m^{-2}). The specific resistance (consider only the electrolyte in this problem) is 2 m Ωm^2, and the Tafel kinetics is given as $\eta = -0.05 + 0.07 \log j$, with current density units of A m^{-2}.

a) Determine the cell potential.
b) Determine the efficiency.

The manufacturer has decided to make a battery with thinner electrodes. That is, the active material is thinner, whereas the supporting current collector remains of the same thickness. The new electrodes have a quarter of the original thickness. The battery package is going to be charged at 4 C.

c) Determine the new current density.
d) Determine the new cell potential.
e) Determine the efficiency.
f) How much heat (in kW) is this for a battery package of 45 kWh?

SOLUTIONS

Solution to Problem 7.1. Battery Characteristics.

a) When charging from 20 to 80% SoC (0.6 change in SoC) in 40 minutes (0.67 h), the C-rate becomes $0.60/0.67 = 0.89C$. Thus the 40 minutes charging is done at an C-rate of 0.89. The current density thus becomes:

$$j_2 = j_1 \frac{C_2}{C_1} = 20 \ [\text{A/m}^2] \frac{0.89}{1} = 17.8 \ [\text{A/m}^2].$$

b) From the figure we can read:

$$E^{ocp} = 2.8 \text{ [V]} \quad \text{and} \quad r = 2.3 \text{ m}\,\Omega\text{m}^2.$$

c) The cell potential becomes

$$E^{cell} = E^{ocp} + rj + \eta$$
$$= 2.8 + 0.0023 \cdot 17.8 + (-0.02 + 0.03\log[17.8]) = 2.86 \text{ [V]}.$$

Solution to Problem 7.2. Long-Range Battery Vehicles.

a) The C-rate becomes $60/10 = 6$ C.

b) The current density becomes

$$j_2 = j_1\frac{C_2}{C_1} = 20 \text{ [A/m}^2\text{]}\frac{6}{1} = 120 \text{ [A/m}^2\text{]}.$$

c)

$$E^{cell} = E^{ocp} + rj + \eta = 3.65 + 0.002 \cdot 120 + (-0.02 + 0.03\log[120])$$
$$= 3.93 \text{ [V]}.$$

d)

$$\varepsilon = \frac{E^{ocp}}{E^{cell}} = \frac{3.65}{3.93} = 93\%.$$

e) The charging power needed is

$$P_{min} = \frac{E_{20 \text{ kW for 5 h}}}{1/6 \text{ [h]}} = \frac{20 \text{ [kW]}5 \text{ [h]}}{0.167 \text{ [h]}} = 0.60 \text{ [MW]}.$$

f) The irreversible heat becomes

$$\dot{Q} = P_{actual} - P_{theor.} = P_{theor.}\left(\frac{1}{\varepsilon} - 1\right)$$
$$= 600 \text{ [kW]}\left(\frac{1}{0.93} - 1\right) = 45 \text{ [kW]}.$$

This is a remarkable amount of heat. Even with very good cooling heating systems, this heat production shows why fast charging like this is a challenge, certainly for a personal vehicle. For buses and equivalents, an advanced cooling system is more relevant. This example shows why phase change elements are considered for heat adsorption like in Problem 4.1.

Solution to Problem 7.3. High-Power Battery Design.

a) The cell potential is given as

$$E^{cell} = E^{ocp} + rj + |\eta|$$
$$= 3.6\ [\text{V}] + 0.002\ [\Omega\text{m}^2]50\ [\text{A}\,\text{m}^{-2}] + |(-0.05 + 0.07\log 50)|\ [\text{V}]$$
$$= 4.02\ [\text{V}].$$

b) The efficiency is given by the ratio between the open and closed cell potentials:

$$\varepsilon = \frac{E^{ocp}}{E^{cell}} = \frac{3.6\ [\text{V}]}{4.02\ [\text{V}]} = 90.0\%.$$

c) This problem is about how to design high-power battery packages. When the electrodes become thinner, the current density across the electrolyte becomes proportionally thinner at constant C-rate. The C-rate is proportional to the current density and thus we get:

$$j_{new} = \frac{C - rate_{new}}{C - rate_{old}} \frac{\delta_{old}}{\delta_{new}} j_{old} = \frac{4}{2}\frac{C}{C}\frac{1}{4}50\ [\text{A/m}^2] = 25\ [\text{A/m}^2].$$

Additionally, the exchange current density will also change. That is, the exchange current density that we observe from a cross-sectional point of view changes proportionally to the electrodes thickness. The new exchange current density is as follows:

$$j_{0,new} = \frac{\delta_{new}}{\delta_{old}} j_{0,old} = 0.25 j_{0,gml}.$$

The coefficient a changes accordingly:

$$a_{new} = -b\ \log j_{0,new} = -b\ \log[0.25 j_{0,old}] = -b\ \log j_{0,old} - b\ \log[0.25]$$
$$= a_{gml} - 0.07(-1) = -0.05 + 0.042 = -0.008\ [\text{V}].$$

d) The cell potential becomes

$$E^{cell} = E^{ocp} + rj + |\eta|$$
$$= 3.6\ [\text{V}] + 0.002\ [\Omega\text{m}^2]25\ [\text{A}\,\text{m}^{-2}] + |(-0.008 + 0.07\log 10)|\ [\text{V}]$$
$$= 3.74\ [\text{V}].$$

e) The efficiency becomes

$$\varepsilon = \frac{E^{ocp}}{E^{cell}} = \frac{3.6}{3.74} = 96.3\%.$$

Remarkably, we double the total input current and, at the same time, improve the efficiency. Lowering the thickness to a quarter is more than what is done in reality; however, lowering the electrode thickness by a quarter (from 100 to 75 µm) is realistic for high-power batteries.

f) The theoretical power needed for the battery to charge in 0.25 h is

$$P_{theory} = \frac{Energy}{time} = \frac{45\ [\text{kWh}]}{0.25\ [\text{h}]} = 180\ \text{kW}.$$

Moreover, 96% of the input energy remains as work, and some more power is needed. The heat becomes the difference between the needed power and the remaining power:

$$\dot{Q} = (P - P_{theor.}) = \left(\frac{P_{theor.}}{\varepsilon} - P_{theor.}\right)$$
$$= \left(\frac{180\ [\text{kW}]}{0.963} - 180\ [\text{kW}]\right) = 6.92\ \text{kW}.$$

This is equivalent to having all three electric plates on a stove going at max power. This demonstrates the importance for heat management when charging Li–ion batteries, even at high efficiency and for short times (15 min).

CHAPTER 8

Hydrogen for Energy Storage

Hydrogen, when used for energy storage, is the most flexible energy storage medium available. Considering a nonfossil fuel-based economy, we can engineer energy storage, distribution, and propulsion systems from less than a kW up to several MW using hydrogen systems. We can design systems that balance energy fluctuations for some minutes up to seasonal variation. This flexibility is why hydrogen as an energy storage medium is so interesting and why its branch of technologies has received so much research and market attention. This chapter evaluates the how and why and gives much engineering details to compare systems quantitatively.

On a molar basis, hydrogen is the most omnipresent element on the surface of this planet. Also, it is the element most strongly bound to oxygen, relative to its own mass. These two properties, omnipresence and supreme specific energy, make it one of the most interesting substances to be used as an energy carrier in the transportation sector.

The natural form of hydrogen in nature is being bound to oxygen in the form of water, H_2O, simply due to the strong affinity between the elements. Oxygen, on the other hand, is present in its elemental form, and this means that if we can get some hydrogen on board of a vehicle, we will have the lightest energy source available. We only need to overcome three defining challenges; i) getting hydrogen in its pure elemental form, ii) getting it on the vehicle in a storable form, and iii) find a way to efficiently convert it into power for vehicles or other applications.

The energy train of hydrogen as an electric energy storage medium is illustrated in Fig. 8.1. It is illustrated by the use of a wind turbine that gives more electricity than necessarily needed. Imagine that a wind park can stay without much production for some days or produce much more than needed during other days. Photovoltaic (PV) electric power units holds similar intermittent properties, only more periodic since sun light is absent at night. This energy can be stored for later use, and hydrogen is an option. Later the hydrogen can be reused for transportation or for electricity and heat in residents, public and office buildings via a combined heat and power route.

Hydrogen as an electric energy storage medium requires; i) hydrogen production by electrolysis, ii) hydrogen storage and distribution, and

Engineering Energy Storage.
DOI: 10.1016/B978-0-12-814100-7.00008-0
Copyright © 2017 Elsevier Inc. All rights reserved.

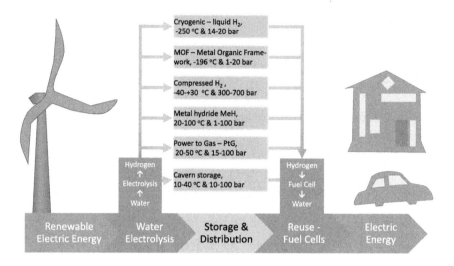

Figure 8.1 Hydrogen as an electric energy storage medium.

iii) reuse via fuel cells. Hydrogen can be made from more than electricity, and currently reformation from natural gas is the most common method. Coal can also be used for reformation into hydrogen. Since this is a book about energy storage and reformation is rather about energy conversion than about energy storage, these two processes are treated very lightly; see Section 8.1.3.

To understand just how convenient hydrogen is as an energy storage medium, we should compare it to other energy storage media used today. Let us compare three variables: specific energy (J/kg), volumetric energy (J/m^3), and system energy capacity (J). The last one is different from the two others since it is an extensive property. It is used to compare the sizeability of the technologies.

In Table 8.1, energy storage media are listed with the three aforementioned energy terms: specific energy, volumetric energy, and total energy of a relevant system unit. The energy storage media are listed based on the amount of energy stored in actual relevant systems. Starting with lithium ion batteries, a large-scale energy system can be a bus with over night charging. A bus like this comes with a battery package of up to 300 kWh. This is the lowest system energy total capacity in this comparison. The energy density is taken from Table 7.3, p. 117. Next is the energy of a compressed air energy storage system, where the by far largest plant currently planned is aimed to be 800 MWh. However, in this comparison, we choose a quarter

Table 8.1 Overview of specific, volumetric, and total system unit energy for selected energy storage media. Values are based on media enthalpy or gravimetric potential energy

Technology	$e/\mathrm{kWh\,kg}^{-1}$	$\hat{e}/\mathrm{MWh\,m}^{-3}$	$E_{Inst.}/\mathrm{MWh}$
Li-ion battery (LCO)	0.19	0.56	0.30[a]
Compr. Air Energy Stor. (CAES)	0.14	0.012	200[b]
Hydrogen, 1 bar	33	0.0027	334[c]
Hydroelectric, pumped, 400 m	0.0011	0.0011	1,500[d]
Hydrogen, 700 bar	33	1.6	2,623[e]
Hydrogen liquid	33	2	135,113[f]
Liquid Natural Gas	15	6.1	352,775[f]
Gasoline	13	9	545,197[f]
Diesel	13	10	577,268[f]
Jet-A fuel	13	10	593,303[f]

[a] For a bus with 300 kWh battery capacity
[b] For a capacity of existing and planned installations
[c] For a 100 km pipeline, with Ø = 0.4 m, and at 10 bar pressure
[d] A quarter of Bath County Pumped Storage Station, charging for 2 hours
[e] For the volume of underground CAES and at 350 bar
[f] For the volume of a typical sized oil tank, $h = 15$ m height and Ø = 70 m

of the largest currently possible. The energy density is from Example 5.1, p. 65. Hydrogen, H_2, will never be stored at one bar. If the society was built up around hydrogen, a pipeline of 100 km, 40 cm in diameter, and at 10 bar would be equal to 334 MWh, about twice to that of CAES. If, for example, an underground cave of the CAES was filled with hydrogen at 350 bar instead of air, the energy in the storage unit would be more than ten times larger than in the CEAS system. This would in fact be almost twice a typical hydroelectric pumping facility (evaluating a quarter of the world's largest hydroelectric pumping facility (6 GW) going for two hours). However, it is liquid fuels that allow for the most energy storage possible. Cylindrical fuel tanks that we can see at air ports and oil refineries are sized in the order of 15 m in height and 70 m in diameter, depending on the content and need. Considering this volume for liquid hydrogen, liquid natural (or bio) gas, gasoline, diesel, and kerosene (Jet-A fuel), we see two new orders of magnitude increase in system unit energy content.

Fig. 8.2 graphically shows the volumetric and specific energy of the chosen technologies. It illustrates that when evaluating specific energy, hydrogen is truly the best option for storing energy and that when considering volumetric energy storage, hydrocarbon chains give a better option. It illustrates that carbon is a good hydrogen carrier.

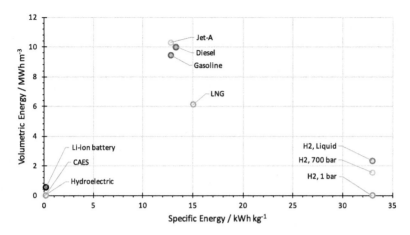

Figure 8.2 Graphical overview of the specific and volumetric energy densities in Table 8.1.

When evaluating the total energy content of different energy systems, given the assumptions in Table 8.1, we see that hydrogen can be used to gap energy storage needs between small-scale and large-scale stationary energy storage. Hydrogen is the lightest energy storage medium available. The facts that hydrogen is suitable for energy storage on so many different levels and that it is highly compatible with renewable energy production explain why it is considered so extremely promising as an energy carrier.

Example 8.1: Hydrogen Stored in Pipelines.

If one would build a gas pipeline between Toronto (Canada) and Vancouver (Canada) with the dimensions as in Table 8.1 but with 50-bar pressure and twice the diameter, how many tons of oil equivalents (toe) would the pipeline contain? Consider a distance of 4350 km.

Solution:

The energy content per 100 km would be five times larger from the pressure and another 4 times because of the diameter doubling, and the length 43.5 times longer, that is, 291 GWh. One GWh is equal to 860 toe, meaning that the energy stored would be 25 ktoe.

Canada has about 35 million inhabitants that each on average use 7.3 toe annually. The daily energy need of the nation is thus around 700 ktoe per day. Thus such a pipeline would contain 3% of the daily energy need. If, on the other hand, considering that the energy fluctuations are from the residential sector and that this in turn is 15–20% of the total consumption, such a pipe line would offer around 20% energy buffer on a daily basis. This shows two things: hydrogen can store a significant amount of our energy usage and that we use a lot of energy.

8.1 HYDROGEN PRODUCTION. WATER ELECTROLYSIS

Overall, hydrogen electrolysis consists of water being split into hydrogen and oxygen using electric work:

$$\text{General: } H_2O \Rightarrow H_2 + \frac{1}{2}O_2, \tag{8.1}$$

$$\text{Reversibly: } H_2O + \text{electric work} \Rightarrow H_2 + \frac{1}{2}O_2 - \text{reversible heat}, \tag{8.2}$$

$$\text{Irreversibly: } H_2O + \text{electric work} \Rightarrow H_2 + \frac{1}{2}O_2. \tag{8.3}$$

The thermodynamic basis of this is that work, equivalent to Gibbs free energy $\Delta \bar{g}$, is required as a minimum input. In addition, we must add work to compensate friction in the form of ohmic resistance in the electrolyzer and Tafel friction from electron transfer at the electrodes. Several technologies to perform electrolysis are developed; the denominator is that they must all pay the energy cost of Gibbs free energy and that beyond this they all have different and characteristic energy coss due to ohmic resistance and Tafel overpotentials.

8.1.1 Water Electrolysis Thermodynamics

The thermodynamics of water electrolysis covers Gibbs free energy of formation $\Delta \bar{g}$, reaction enthalpy $\Delta \bar{h}$, and reversible heat $T\Delta \bar{s}$ as the primary overall energies. Additionally, the half-cell reversible potentials are included as more detailed and secondary information.

8.1.1.1 The Energies

The work from a spontaneous process changes with temperature. Rewriting Eq. (2.34), p. 25, by saying that $\Delta \bar{s} = \Delta \bar{s}^o + \bar{R}\ln\frac{p}{p^o}$ or $p = p^o$, we get Eq. (8.4) or (8.5):

$$\Delta \bar{g}^o = \Delta \bar{h} - T\Delta \bar{s}^o, \tag{8.4}$$

$$\Delta \bar{g} = \Delta \bar{h} - T\Delta \bar{s}. \tag{8.5}$$

A common simplification that we will use here is that the reaction enthalpy and entropy are constant with temperature. This is a simplification that will

Table 8.2 Thermodynamic data for the formation of liquid and gaseous water from hydrogen and oxygen at 1 bar and 20°C [2]

Reaksjon	$\Delta \bar{h}^0/\text{kJ mol}^{-1}$	$\Delta \bar{g}^0/\text{kJ mol}^{-1}$	$\Delta \bar{s}^0/\text{J mol}^{-1}$
$H_{2,(g)} + \frac{1}{2}O_{2,(l)} \Rightarrow H_2O_{(l)}$	-286	-237	-167
$H_{2,(g)} + \frac{1}{2}O_{2,(g)} \Rightarrow H_2O_{(l)}$	-242	-229	-44

not hold for detailed engineering examples. However, relative to their own values, the change is small, and the assumption is often considered valid. Considering the level of details in this book, we use this simplification.

Looking at Eq. (8.1), it is not clear what state the water is in: is it solid, liquid, or vapor? In Table 8.2, the thermodynamic data for the reaction energies and entropies are given. When considering one bar, we can suppose that below 100°C, the values involving liquid water are relevant, and above 100°C, the values correspond to vapor.

Example 8.2: Reversible Potential at Different Temperatures.

What is the standard potential and reversible heat for the hydrogen reaction at 90 and 110°C, respectively, with a pressure of 1 bar.

Solution:

From Table 8.2 we can calculate the standard Gibbs free molar energy for each of the temperatures under the assumption that the reaction entropy and the enthalpy do not change with temperature. From Eq. (8.4) we get the Gibbs molar energy, and from the Nernst equation (using 2 equivalents per mole hydrogen), we get the standard potential:

Temperature	$\Delta \bar{g}^0/\text{kJ mol}^{-1}$	E^0/V	$\bar{q}_{rev.}$
90°C	-225	1.17	-61
110°C	-225	1.17	-17

This example shows that the difference between the standard potential between the reactions is very similar around 100°C, whereas the reversible heat changes a lot. This is due to the phase change of water.

8.1.1.2 Half Cell Potentials and pH

When looking at the tabulated standard half-cell potentials at room temperature, we find that the standard potential for the hydrogen reduction reaction in 1.0 M H_{aq}^+ is 0 V. This potential and electrode setup is the standard reference for all the other reactions. When bubbling hydrogen over a platinum electrode in a solution of, for example, 1 M HCl at 1 bar, we term this the standard hydrogen electrode and define this as the standard hydrogen electrode voltage V_{SHE}. This reaction is given in Eq. (8.6).

From reaction (8.6) we can calculate the standard half-cell potential of the oxygen reduction reaction (ORR) in this standard acidic solution as well. From Table 8.2 we have that the total standard potential at room temperature is 1.23 V (from $\frac{237\ 000\ [J/\ molK]}{2 \cdot 96485\ [C/mol]}$). The half cell then becomes 1.23 V_{SHE} (Eq. (8.7)).

$$2H^{+}_{(aq.)} + 2e^{-} \rightarrow H_{2.(g)}, \quad E^{o}_{H^{+}/H_2} = 0 \ V_{SHE}, \quad pH = 0, \quad (8.6)$$

$$\frac{1}{2}O_{2.(g)} + 2H^{+}_{(aq.)} + 2e^{-} \rightarrow H_2O_{(l)}, \quad E^{o}_{H_2O/O_2} = 1.23 \ V_{SHE}, \quad pH = 0.$$
$$(8.7)$$

The standard potential for the hydrogen reduction reaction (HRR) is defined when the concentration of protons is 1 M. The definition of pH is the negative Briggsian logarithm of the concentration of protons:

$$pH = -\log\left[H^{+}_{(aq.)}\right]. \quad (8.8)$$

The reversible potential of the HRR described in Eq. (8.6) is defined as

$$E^{rev}_{H^{+}/H_2} = E^{o}_{H^{+}/H_2} - \frac{RT}{zF} \ln \frac{(C^{o}_{H^{+}})^2 P_{H_2}}{(C_{H^{+}})^2 P^{o}_{H_2}}. \quad (8.9)$$

With $C^{o}_{H^{+}} = 1$, $P_{H_2} = P^{o}_{H_2}$, $T = 293$ K, using $\ln[x] = 2.303 \log[x]$, $\log[x^n] = n\log[x]$, and $\log[1/x] = -\log[x] = px$, we obtain

$$E^{rev}_{H^{+}/H_2} = 0 - 2.303 \frac{8.314 \ [J/mol\,K] \cdot 298 \ [K]}{2 \cdot 96485 \ [C/mol]} \log \frac{1}{(C_{H^{+}})^2}$$
$$= -0.059 \ pH. \quad (8.10)$$

Likewise, we can do the same for the oxygen reduction reaction (ORR), as in Eq. (8.7):

$$E^{rev}_{O_2/H_2O} = 1.23 - \frac{RT}{zF} \ln \frac{(C^{o}_{H^{+}})^2 P^{\frac{1}{2}}_{O_2}}{(C_{H^{+}})^2 (P^{o}_{O_2})^{\frac{1}{2}}} = 1.23 - 0.059 \ pH. \quad (8.11)$$

The half-cell potentials lower with increasing pH. For 1 M caustic soda (NaOH), the pH is 14. The half-cell potentials of the HRR and ORR become -0.40 V_{SHE} and 0.83 V_{SHE}, respectively. Under this condition,

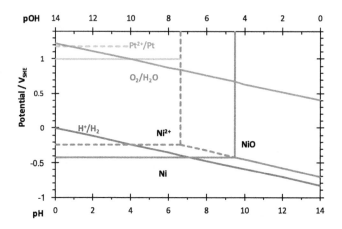

Figure 8.3 Pourbaix diagram for the relevant electrode reduction reactions in water electrolysis and potential electrode materials.

the concentration of protons versus hydroxide ions is so low that the reactions are considered with hydroxide ions rather than protons. Thus the HRR in Eq. (8.6) becomes Eq. (8.12), and the ORR in Eq. (8.7) becomes Eq. (8.13):

$$H_2O_{(l)} + 2e^- \rightarrow H_{2,(g)} + 2OH^-_{(aq.)}, \quad E^0_{H_2O/H_2} = 0.83 \text{ V}_{SHE}, \quad pH = 14, \tag{8.12}$$

$$\frac{1}{2}O_{2,(g)} + H_2O_{(l)} + 2e^- \rightarrow 2OH^-_{(aq.)}, \quad E^0_{O_2/OH^-} = -0.40 \text{ V}_{SHE}, \quad pH = 14. \tag{8.13}$$

The way the half-cell potentials change with pH is shown in Fig. 8.3, where the blue (dark gray in print versions) line shows the HRR, and the green (mid gray in print versions) line the ORR. Additionally, the figure also shows the standard (dashed) and reversible (solid) reduction potentials of two commonly used electrode materials in hydrogen conversion systems, platinum (yellow (light gray in print versions)) and nickel (gray). The diagram in Fig. 8.3 is called a Pourbaix diagram. As the different aqueous water electrolysis technologies are introduced, we shall return to this diagram.

8.1.2 Electrolysis Technologies

Electrolysis is technology that is a hundred years old. As an example, Norsk Hydro started large-scale electrolysis late in the 1920s and has performed

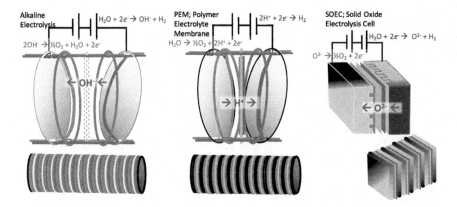

Figure 8.4 Sketch of the three dominating electrolysis technologies: Alkaline, PEM, and SOEC. Green (mid gray in print versions) indicate flow paths removing produced oxygen, and blue (dark gray in print versions) flow paths removing produced hydrogen. Single cells are illustrated in the upper part and stack configurations in the lower part.

megawatt-scale water electrolysis for more than 50 years. This illustrates that, as an industrial technology, water electrolysis is extremely well known and established. On a commercial level, today, three technologies dominate: alkaline electrolysis, polymer electrolyte membrane (PEM) electrolysis, and solid oxide electrolysis cells (SOEC). These technologies are commercially available, however, differently matured.

8.1.2.1 Alkaline Water Electrolysis

When running an electrolysis cell, the anode overpotential increases the potential above the reversible potential. In 1 M HCl, the half-cell potential is 1.23 V_{SHE}, and only gold will not oxidize and dissolve at such potential. This is why, when initially developing commercial water electrolysis 100 years ago, one opted for a high pH solution, where the anode potential is lower.

Evaluating nickel as a cathode material, the nickel reduction reaction (NRR) must be above the reaction it competes with to avoid oxidation. At low pH, the oxidized nickel dissolves, whereas at higher pH, it forms a stable oxide (NiO) layer. This is indicated in Fig. 8.3. Moreover, the HRR potential depends on the concentration of ions in solution. 1 M nickel gives the standard potential; however, this is not the case. The amount of nickel in the solution is not necessarily given, but if considering a concentration of 0.001 M, then the reversible potential and dissolution lines of nickel are given by the dashed gray lines. We now see that, in an alkaline solution, the HRR cathode reaction stabilizes nickel and the ORR anode reaction stabilizes nickel oxide. These are the premises for material selection in alkaline water electrolysis.

The cell design of a water electrolysis system consists of unit cells stacked together. The single unit cells can have different configurations, and a common one is illustrated to the upper left in Fig. 8.4. This setup utilizes bipolar plates that produce hydrogen on one side and oxygen on the other side. In the illustration, this is shown from a slightly different perspective. To the very left in the alkaline electrolysis unit cell, we see the anode reaction taking place. Hydroxide ions (OH^-) are consumed, and oxygen and water are produced. The hydroxide ions are in turn produced at the cathode along with hydrogen. These ions migrate from the cathode to the anode. When the gas bubbles are produced, they must be prevented from mixing (H_2-bubbles vs O_2-bubbles). This is done by inserting a mechanical barrier, a diaphragm. The diaphragm is a barrier not only for gas bubbles, but also for the ionic current. The diaphragm must therefore be as porous as possible and as little tortuous as possible while simultaneously prevent the gas bubbles from passing over. As gas bubbles typically are larger than 10 μm, the pore size is smaller [47]. The diaphragm is in the order of 10-mm thick and has a specific ohmic resistance of around 1 $\Omega\,cm^2$ [47]. Between the diaphragm and the electrodes, water flows, or more specifically, potassium hydroxide (KOH) is saturated with oxygen in the anolyte, and hydrogen in the catholyte. Anolyte is the electrolyte on the anode side of the diaphragm, and catholyte is the electrolyte on the cathode side. Preventing gas bubbles of hydrogen and oxygen from mixing is as much a safety aspect as an economical one. Mixing the two gasses leads to a back reaction to water and thus lost work and loss in galvanic efficiency. Galvanic efficiency is essentially the number of moles of hydrogen that is produced relative to the number of coulombs (moles of electrons) that passes the cell. To prevent mixing, the pressure of the anolyte and catholyte on each side of the diaphragm must be the same at any point; otherwise, oxygen saturated anolyte would flow across the diaphragm into the catholyte or vice versa. To lower the ohmic resistance as much as possible, focus is also kept on lowering the distance between the diaphragm and each of the electrodes. There are several different approaches in this effort: one is to apply a porous mesh adjacent to the diaphragm similar to the PEM design (see the next section), the second is to apply active nodes that produce gases only very close to the diaphragm (more like alkaline sketch in Fig. 8.4), and the third

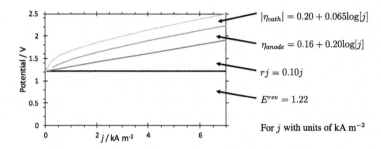

Figure 8.5 Energy input in an example alkaline water electrolysis with values taken from [47].

is to have a plate and open space between the diaphragm and the electrode area (most like the sketch in Fig. 8.4). More alternatives exist, consisting of hybrids of these three strategies. A more compact design with porous active nodes is good for the ohmic resistance of the cell on the one hand, however much more challenging for the fluid mechanics on the other because of the electrolyte flow pressure drop must be the same on each side of the diaphragm to prevent cross flow and efficiency loss.

Electrolysis unit cells are stacked, most typically, in series. This is done similarly to what is done with disposable batteries in consumer electronics, however, with one major difference. Since both the anode and the cathode consist of nickel, one side becomes the anode with a nickel oxide layer, and the other becomes the cathode. The bulk metal becomes the connector between the anode and the cathode of two unit cells, cast into one component. When a plate like this is anode in one cell and the cathode of the neighboring cell, one side is polarized in the cathodic direction, and the other is polarized in the anodic direction. Such a plate is termed a bipolar plate. Bipolar plates are very common in several electrochemical systems because they make the system more compact and give a simpler (fewer components) design.

The energy consumption of alkaline water electrolyzers vary by design, but foremost by current density. At low current density, the electron transfer process represents the main loss in energy efficiency, whereas at larger current density, the ohmic current density dominates. This is shown in Fig. 8.5. The lower horizontal line is the reversible work required, E^{rev}. On top of this, there come the energy losses: the ohmic potential rj, anode overpotential η_{anode}, and the absolute value of the cathode overpotential $|\eta_{cath}|$. This information, as shown in Example 8.3, can be used to find the energy need per kg hydrogen, the energy efficiency, and the sources of the energy efficiency at different current densities.

Example 8.3: Energy Consumption and Efficiency in Electrolysis.

Using the information in Fig. 8.5, determine the energy per kg hydrogen, the energy efficiency, and the relative contribution to the efficiency loss at current density of 0.5 and 4 kA m^{-2}

Solution:

The energy need per mole is given by the cell potential and the Faraday constant and equivalents per mole hydrogen (2 coulombs or moles of electrons per mole of hydrogen). Dividing by the molar weight (0.002 kg mole^{-2}) gives the specific energy:

$$e_{H_2} = \frac{\bar{e}_{H_2}}{M_{H_2}} = \frac{zFE^{cell}}{M_{H_2}} = \frac{zF(E^{rev} + rj + \eta_{anode} + |\eta_{ccath}|)}{M_{H_2}}.$$

Energy efficiency is given by the energy conserved versus the energy that goes into the system:

$$\varepsilon = \frac{E^{rev}}{E^{rev} + rj + \eta_{anode} + |\eta_{ccath}|}.$$

The relative contributions is given as

$$f_{rj} = \frac{rj}{E^{cell}}, f_{\eta_{anode}} = \frac{\eta_{anode}}{E^{cell}}, and f_{\eta_{cath}} = \frac{|\eta_{cath}|}{E^{cell}}.$$

Thus the results follows:

j /kA m^{-2}	E^{cell} /V	e_{H_2}-need /kWh kg$^{-1}_{H_2}$	ε %	f_{rj} %	$f_{\eta_{anode}}$ %	$f_{\eta_{cath}}$ %
0.5	1.55	42	79	15	30	55
4.0	2.13	57	57	43	31	26

We can see that the energy efficiency lowers significantly when the current density increases. We can also see that the main contribution to the loss in energy efficiency at high conversion rate is the ohmic potential loss. Therefore, to produce hydrogen more intensively and more efficiently, we should focus on lowering the ohmic resistance of the system.

8.1.2.2 PEM Water Electrolysis

The defining part of a PEM water electrolyzer is the polymer electrolyte membrane (PEM). Compared to the alkaline electrolyzer that uses a diaphragm of around 10 mm, a PEM water electrolyzer uses a membrane of around 0.05 mm (in fact, 1–3 milliinches). The electrodes are thin porous layers (0.002–0.010-mm thick) and are layered on each side of the membrane, constituting a membrane electrode assembly (MEA). The MEA is the core of a PEM electrolyzer, and the gas bubbles come out at each side of the MEA and are removed by pH neutral water flowing on each side. The concept of a PEM water electrolysis cell single unit and stack is sketch in the middle of Fig. 8.4.

The most commonly used membrane is Nafion. This membrane is polymeric with a poly-tetra-fluor-ethylene (PTFE) like structure containing functional groups that create micro- or nanochannels that allow protons and some water to pass through. The conductivity for wet membrane well above the room temperature (60–90°C) is in the order of 10 $S m^{-1}$, [48] giving a specific resistance of around 0.005 $m\Omega m^{-2}$. This very low ohmic resistance of the cell is one of the advantages of the PEM technology.

A second advantage for the PEM water electrolysis is the kinetics for hydrogen evolution reaction (HER) in acidic solutions. The charge transfer overpotential (Tafel) is nearly absent and can even be considered to be constant of order 1–10 mV, depending on the catalyst loading.

The main challenge with PEM water electrolysis is the conditions for the anode. Looking at the Pourbaix diagram in Fig. 8.3, we can see that, for instance, platinum will actively dissolve when the potential goes beyond 1–1.2 V_{SHE} and at a pH of 0–2. Gold can withstand higher potentials but has no very good catalytic properties (large Tafel potential). The solution is then using a stable metal oxide, more specifically, an oxide based on ruthenium, RuO_2. Ruthenium oxide in the right form (rutile) has a very high electronic conductivity for being a metal oxide. Ruthenium is a rare earth metal and very expensive. Several efforts are made to modify the ruthenium oxide by mixing it with similar elements like iridium, rhodium, or osmium, and this represents in part of the research forefront in this area [49].

When considering energy efficiency of a PEM water electrolyzer, the cell potential as a function of conversion rate (usually expressed as j) is presented in Fig. 8.6. We can see that the largest contribution to lost work is the anode charge transfer overpotential. Performing the exercise of Example 8.3, however for 1 and 30 $kA m^{-2}$, we find energy efficiencies of 51 and 54% percent, respectively. The relative contribution to the loss in efficiency from the anode Tafel overpotential is 98 and 78% for the two current densities. Lowering the anode charge transfer overpotential is thus a key point, and this illustrates in part why PEM water electrolysis is challenging to develop and developed after the alkaline electrolyzer. On the other hand, we can also see that the possible conversion rate is tenfold to

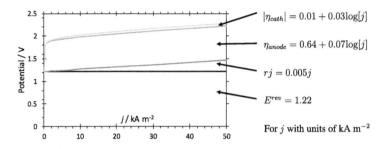

Figure 8.6 Energy input in an example PEM water electrolyzer with values taken from [48,49].

Table 8.3 Specific electric energy consumption and energy efficiency for the three electrolysis concepts of alkaline electrolysis, PEM electrolysis using Nafion, and alkaline PEM water electrolysis

Rate	Alkaline		PEM		Alk. PEM	
j kA m^{-2}	e_{H_2}/kWh kg^{-1}	$\varepsilon\%$	e_{H_2}/kWh kg^{-1}	$\varepsilon\%$	e_{H_2}/kWh kg^{-1}	$\varepsilon\%$
1	45	73	51	64	45	73
4	57	57	53	61	49	66
40	160	20	54	60	51	64

that of the alkaline electrolyzer and still at reasonable energy efficiencies; PEM water electrolysis offers compact units with very high production rate and at reasonable energy efficiency, as can be seen from Table 8.3.

The sketch of the PEM electrolyzer in Fig. 8.4 consider water flowing through to remove the products, hydrogen and oxygen. One possible design is to have only gas on the hydrogen side since the need for water is only on the anode side. Thus hydrogen gas streams out of the cell. When assembling the PEM water electrolysis cell into a stack, the concept of bipolar plates is introduced once again. This means that the anode side of the bipolar plate is polarized below the HRR potential and the cathode side above the ORR potential. Even though the anode side is in contact with pure water, the potential ends up much higher than the 1 V indicated because of the additional charge transfer overpotential, typically, additional 0.5–0.8V. As said, very few materials can sustain polarization at these potentials without undergoing oxidation, leading to a passivated electrically insulating oxide layer or active oxidation dissolution. Therefore, another main research challenge in the PEM water electrolysis development consists of finding protective layers for this side of the bipolar plate.

PEM is commonly considered an acronym for polymer electrolyte membrane; however, proton exchange membrane is also rightfully used. Under the first consideration (polymer rather than proton), we could consider membranes that conduct hydroxides rather than protons. The challenge has been, however, that the only membrane that can sustain temperature over time has been Nafion, and this membrane conducts protons rather than hydroxide ions. However, recently, hydrocarbon membranes that can sustain hydroxide solutions for very long times and with similar conductivity as Nafion have been recently developed [50]. This allows for the usage of the catalysts of the alkaline electrolyzer and the resistance of the PEM electrolyzer. This would lead to improved efficiency compared to the alkaline electrolyzer and the PEM electrolyzer using Nafion and also cheaper electrode materials compared to a "traditional" PEM water electrolyzer. This example is however still on a research level and not yet on the market. Fig. 8.6.

8.1.2.3 Solid Oxide Electrolysis Cells

Solid oxide electrolysis cells (SOEC) are a third electrolysis technology characterized by having membrane that consists of a solid mixed metal oxide, where the oxygen ions O^{2-} are transported through the membrane.

The metal oxide is based on a metal that binds to oxygen via four electron clouds as a metal dioxide, MeO_2. This way, a solid network where each metal atom is connected to other metals indirectly via oxygen ions is formed. By replacing around ten percent of the metal with another similar metal that has the property of binding only to three oxygen atoms, oxygen vacancies are introduced in the network structure. At sufficiently high temperatures, the remaining oxygen ions become mobile and start to move around in the network. This happens when an oxygen ion next to a vacancy moves into the vacancy, creating a new vacancy in its former spot. In this way, oxygen ions can move and, when polarized, migrate. The most commonly studied SOEC membrane consists of yttria stabilized zirconia (YSZ), $Zr_{0.9}Y_{0.1}O_2$ [51,52], where zirconia is the base metal oxide in the structure, and yttria is the substitute metal oxide that creates the vacancies. Gadolinium-doped cerium oxide (GCD) $Cr_{0.9}Ga_{0.1}O_2$ is another example and also more complex mixed metal oxides like, for example, ceria scandia stabilized zirconia (CSSZ) [53]. The membranes have thickness of order 0.02–0.40 mm and conductivities of order 1–10 $S\,m^{-1}$ [51–53]. The conductivity of these materials follows an Arrhenius behavior (see Eq. (6.26), p. 84), and whereas YSZ has a temperature range relevant for operation above $\sim 700°C$ [51], the CSSZ has a relevant conductivity from $\sim 550°C$ [53].

The electrode materials are typically nickel or platinum at the cathode and nickel oxide mixed with the membrane material at the anode, similar to the alkaline water electrolysis. The difference is that the reactants and the products are all in a gas phase. water vapour enters on the cathode side, and on the anode side, pure oxygen comes out. On the anode side, a mixture of steam and hydrogen comes out, typically, in the order of 90% hydrogen. Water is separated when cooling the outlet gases.

The cell can be designed in a stack with flow channels like the example in Fig. 8.4 (right), where single cells are stacked, but other designs also exist. A common design is actually a tube reactor, where the anode is on the inside, and the cathode on the outside. In such a reactor, the hydrogen and vapor circulate around the tubes, and oxygen comes out at the end of the tubes. The challenge with the SOEC technology is that one must find materials that can withstand the high temperature and also tolerate to be cycled in temperature if the electrolysis cell is not in constant use. The higher the temperature, the greater the material challenge. This is the main motivation behind developing low-temperature (500–600°C) SOEC materials.

When evaluating energy efficiency of water electrolysis, we typically define it as the electric work into the process relative to the work available from the hydrogen afterwards. If doing so with the SOEC, we can get energy efficiencies above 100%. This is realtes to thermodynamics and is not so when accounting for heat intput in addition to electric work input. The reference work output of hydrogen relates to room temperature, and this refers to 1.23 V or $31.8 \text{ kWh kg}_{H_2}^{-1}$. Because of the reversible heat need of the reaction in water electrolysis, the reversible potential decreases as the temperature increases. This need for heat in the process means that, under reversible conditions, it is endothermic and, unless heat is added, the process will cool itself. When cooling, the conductivity lowers exponentially with temperature leading to ohmic heating. Eventually, at sufficiently low temperature, the heat from the ohmic resistance of the cell ηj^2 equals the reversible heat $\frac{T\Delta\bar{s}}{zF}j = -\eta j^2$, and the process becomes adiabatic or thermoneutral. The cell potential at this point is defined by the reaction enthalpy and is referred to as the thermoneutral cell potential E^{TN}. These relations are given by the equation

$$E^{rev} + \eta j = E^{TN} = -\frac{\Delta\bar{h}}{zF} = -\frac{\Delta\bar{g}}{zF} - \frac{T\Delta\bar{s}}{zF}. \tag{8.14}$$

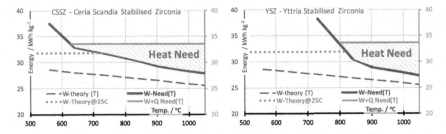

Figure 8.7 Modeled SOEC energy need as a function of different temperatures at 4 kA m^{-2} and for a 0.2 mm thick membrane using CSSZ (left) and YSZ (right) membranes.

In Fig. 8.7, the thermodynamic entities and energy needs for different temperatures are modeled assuming that entropy and enthalpy do not change with temperature. The model accounts for membrane thicknesses of 0.20 mm and current densities of 4 kA m^{-2}; however, thicknesses of a tenth and much larger current densities are possible as shown in the problem section. The left figure evaluates a CSSZ membrane material, and the right figure a YSZ membrane material. The green (light gray in print versions) dotted line gives the work potential at room temperature, $e_{H_2, 298\ K}$, the blue (dark gray in print versions) dashed line the electric reversible energy need as a function of temperature, $e(T)_{H_2}$, the solid blue line the total work needed to run the electrolysis, $e(T)_{H_2} + r(T)j$, and the solid green (light gray in print versions) line the enthalpic work. The intersection point between the solid blue (dark gray in print versions) line and the solid green (light gray in print versions) line gives the temperature where the cell is operated at the thermoneutral cell potential. Above this temperature, there is a heat need because the system is endothermic and below, and cooling is needed because the system is exothermic. If one has access to cheap high-quality heat, for example, from metal winning or nuclear energy, one can overheat steam upon injection and heat the cell so much that the temperature goes above the intersection point between the blue (dark gray in print versions) solid line and the green (light gray in print versions) dotted line. Then, hydrogen can be produced with an electric input that is lower than the output potential value, and from a work analysis point of view, energy efficiencies beyond 100% are possible. For instance, when large surplus of electric energy is available in the grid, then nuclear facilities can produce hydrogen very efficiently. From a thermodynamic point of view, heat and work contributes into the energy efficiency, and in this perspective, the thermoneutral potential is the best case, which is 94% ($\Delta \bar{g}/\Delta \bar{h}$) when considering gas phase SOEC.

8.1.2.4 Other Types of Electrolysis

Going up in temperature significantly lowers the charge transfer over-potentials and to some extent completely remove them. In this respect, high-temperature electrolysis is very interesting. At the same time, one wants to avoid going too high up in temperature because of material thermal degradation challenges. Two remaining technologies are represented in this temperature regime between 100 and 500°C, high-temperature proton conductivity (HTPC) membranes and molten carbonate electrolysis cells (MCEC). These technologies are however not yet commercially available and only discussed as a principle here.

The HTPC systems operate in the range of 200–500°C. Steam is injected on the anode side, and oxygen and steam comes out of the anode side. On the cathode side, pure hydrogen comes out. The membrane conducts protons by a vacancy effect similar to that of solid oxide membrane, only with protons moving from a lattice to a vacancy [54]. The electrode reactions are otherwise the same as shown for the PEM electrolysis in Fig. 8.4. Efficient production, pure hydrogen product (not even water), and relatively simple material engineering aspects are two important benefits for the technology. The main challenge, however, is that in the temperature range of above 200°C, the conductivity of available proton conductors are relatively low, that is, ranging 0.01–0.1 S m^{-1} from 200–300°C [55].

MCEC uses carbonate ions for charge transfer. This complicates the composition of the feed gases to some extent because a mixture of water and carbon dioxide must be fed on the cathode side. The carbon dioxide will in the reaction take the oxygen from water and form a carbonate ion CO_3^{2-}. The carbonate ion migrates through the membrane and over to the anode, where it is oxidized into oxygen and CO_2. The membrane can consist of a porous solid containing a tertiary eutectic mixture of alkaline (Li, Na, and K) carbonate, and the operation temperature can be as low as 397°C [56].

$$\text{Anode:} \quad CO_3^{2-} \Rightarrow 2e^- + CO_{2,(g)} + \frac{1}{2}O_{2,(g)}, \qquad (8.15)$$

$$\text{Cathode:} \quad H_2O_{(g)} + CO_{2,(g)} + 2e^- \Rightarrow H_{2,(g)} + CO_3^{2-}. \qquad (8.16)$$

8.1.3 Hydrogen from Coal and Natural Gas

Currently, about 90% of all the hydrogen is made by reforming carbon and natural gas. This is because the current drive for hydrogen production is the need for hydrogen and not for energy storage. This is changing as more intermittent electric renewable energy (wind and solar PV) is emerging. Currently, most of the hydrogen produced is used at oil refineries to stream line the oil products like diesel gasoline and kerosene (Jet-A), and the easiest way to get hydrogen is to reform hydrocarbons. In the future, when the driver in the hydrogen market is large-scale long-term energy storage, hydrogen production by electrolysis is the solution. Hydrogen by electrolysis meets the market request of handling surplus electric energy. Hydrogen by reforming fossil fuels meets the market requirement of producing cheap hydrogen.

A very simple way to explain natural gas reforming is by feeding natural gas and steam into a reactor that catalyzes hydrogen production. As can be seen from Eq. (8.17), the output is hydrogen and carbon dioxide. By separation, carbon dioxide can be captured and stored underground, but this is yet to be commercially fulfilled (beyond the scope of increasing oil extraction).

$$CH_{4,(g)} + H_2O_{(g)} \Rightarrow 4H_{2,(g)} + CO_{2,(g)}. \tag{8.17}$$

The simple explanation of coal reforming is that carbon and steam are burnt together to give hydrogen and carbon dioxide (see Eq. (8.18)). This is actually more complicated, and carbon $C_{(s)}$ first needs to be gasified. The energy-efficient way to do this is to burn coal with carbon dioxide to form carbon monoxide as a syngas (Eq. (8.19)) and in turn run a shift reaction between steam and the syngas to form hydrogen and carbon dioxide (Eq. (8.20)). Separating the hydrogen leaves the process with carbon dioxide, which in turn is partly stored and partly used to gasify coal.

$$C + H_2O \Rightarrow 2H_2 + CO_2, \tag{8.18}$$

$$\text{Gasification: } CO_{2,(g)} + C_{(s)} \Rightarrow 2CO_{(g)}, \tag{8.19}$$

$$\text{Shift reaction: } CO_{(g)} + H_2O_{(g)} \Rightarrow CO_{2,(g)} + H_{2,(g)}. \tag{8.20}$$

8.2 HYDROGEN STORAGE AND DISTRIBUTION

Hydrogen in the energy sector is for energy storage. Thus hydrogen must be stored. Hydrogen can be produced on site by electrolysis in regions where sufficient electric energy is available. Otherwise, hydrogen must be distributed or transported by other means. In this section, we evaluate the relevant technologies for storage, distribution, and transport, ordered somewhat by relevance and somewhat by maturity.

8.2.1 Thermodynamic Properties of Hydrogen

When evaluating hydrogen storage and distribution, thermodynamics must be dealt with. Three important properties that have a major impact on hydrogen technology are compressibility, phase diagram properties, and spin configuration.

8.2.1.1 Compressibility

When compressing a gas at normal pressure, it is often considered ideal. In an ideal gas, the forces between the molecules are so week that the interaction does not impact the pressure, volume, and temperature relation given by the ideal gas law (Eq. (2.17)). However, as the gas becomes denser, inter molecular forces of contraction and rejection become significant. The equation of state that is used as a first effort to account for these interactions is very often the viral equation. For hydrogen at 300 K, it is given by Eq. (8.21) [57].

$$p = \frac{\bar{R}T}{\bar{\nu}} + \frac{1.438 \cdot 10^{-5} \, [\text{m}^3 \, \text{mol}^{-1}] \bar{R}T}{\bar{\nu}^2} + \frac{3.438 \cdot 10^{-10} \, [\text{m}^6 \, \text{mol}^{-2}] \bar{R}T}{\bar{\nu}^3},$$

(8.21)

where p is pressure in Pa, $\bar{R}$ is the molar gas constant, T is temperature in K, and $\bar{\nu}$ is molar volume in $\text{m}^3 \, \text{mole}^{-1}$. The first term on the right side is recognized as the ideal gas law. The constants in the second and third terms are the correction factors for nonideality and the constant changes with temperature and a lot when the temperature becomes 100 K and lower.

8.2.1.2 Phase Properties

When cooling hydrogen sufficiently, it condenses to a liquid and ultimately freezes to a solid. At 1 K and below, hydrogen has a face center cubic crystal structure, and above 5 K, it has a hexagonal centered packing structure. The triple point is at 0.0695 atm and 13.8 K. The boiling point changes with pressure and temperature, and hydrogen becomes supercritical above

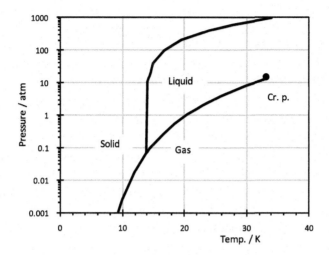

Figure 8.8 Phase diagram of hydrogen as a function of pressure and temperature. Data from [57].

13.8 atm and 33.2 K. The pressure temperature phase diagram is given in Fig. 8.8.

8.2.1.3 Para- and Ortho-Hydrogen

Hydrogen comes in two configurations, depending on the spin of the protons. If the spin of the proton in both hydrogen atoms is the same, then it is termed ortho–hydrogen. If the spins have the opposite direction, then it is termed para–hydrogen. At room temperature and above, hydrogen consists of around 25% para–hydrogen and 75% ortho–hydrogen. This is the equilibrium composition at room temperature and is termed normal hydrogen. The equilibrium content of para–hydrogen at different temperatures is tabulated in Appendix C.

In liquid hydrogen, all hydrogen is para–hydrogen. This means that all ortho–hydrogen must be converted into para–hydrogen. This requires energy removal. The conversion enthalpy of converting normal hydrogen into para–hydrogen at different temperatures is also tabulated in Appendix C.

The spin of hydrogen is conserved in time. This means that if liquid hydrogen is evaporated to room temperature, the gas will consist of pure para–hydrogen. Unless catalyzed, it will take years before converting it to normal hydrogen. In the liquefaction process of hydrogen, special catalyst beds are used to convert normal hydrogen made above 300 K into the composition at the low temperature. More specifically, the catalyst is there to convert ortho-hydrogen to para-hydrogen during cooling.

8.2.2 Hydrogen Storage Technologies

Hydrogen storage technology has been around and been developed for several decades already. Since the PEM fuel cell (see the next section) development accelerated in the 1990s, solutions have been made ready for hydrogen storage and distribution. In the early 2000s, the market was not yet developed, and an overall common strategy was not agreed on. In this environment, all technological solutions had the opportunity to establish and develop. The context of developing these solutions was the hydrogen society, where hydrogen is envisioned to be the main fuel of our society. Here we briefly describe the most relevant solutions.

8.2.2.1 Power to Gas

Investment in renewable energy and especially wind and photovoltaic (PV) electricity production has grown enormously over the last years. If not accounting for public subsidies, the investment into renewable energy production exceeded the investment into nonrenewable sometime around 2014. A lot of this investment was due to political initiatives and PVs becoming financially viable in many places. This development in renewable energy investment came suddenly compared to the electric grid development and market fiscal mechanisms, in turn leading to regions with so large surplus of electric energy that consumers where paid to spend electric energy. These negative prices in electric energy peak periods lead to investment in electrolysis units that would act as electric energy dumps where the electric power was converted into hydrogen gas, and hence the name power-to-gas.

The regions with this surplus of electric power had very scarce markets for the hydrogen, except blending it into the natural gas network. This increases the specific heating value of the gas and also makes it partly renewable. The limitation for feeding hydrogen into natural gas network is what the technology can handle. Hydrogen fragility and gas burners both tolerate up to 20 mol% hydrogen, although most countries put the threshold a lot lower to be on the safe side and more.

A final remark is that pure hydrogen has a much higher value than hydrogen with impurities, and power-to-gas in its current form should be seen as a first effort to meet the demand for energy storage rather than the final solution in a hydrogen economy. Nevertheless, this is the first example of commercial large-scale energy storage that has come as a response to the investment in intermittent renewable energy production technology (wind and PV).

8.2.2.2 Compressed Hydrogen

When considering hydrogen for cars, hydrogen compressed to 700 bar (in some instances, only 350 bar) is generally considered the most convenient technology. This is as it meets the requirement for driving range (> 450 km) and sufficiently low fueling time (< 3 min).

Compressing hydrogen requires work. The way this is done is by using a series of compressors and a stepwise compression up to the desired compression. The stepwise compression means that the compression is close to isothermal and more efficient [58]. The specific work for compression of hydrogen can therefore be calculated by integrating pressure with respect to volume (Eq. (2.13)). By inserting the viral equation of state, Eq. (8.21), at constant temperature, we obtain the molar work

$$
\begin{aligned}
d\bar{w}_{H_2} &= \int_1^2 p \, d\bar{v} \\
&= \bar{R}T \int_1^2 \left(\frac{1}{\bar{v}} + \frac{1.438 \cdot 10^{-5} \, [\mathrm{m^3\,mol^{-1}}]}{\bar{v}^2} \right. \\
&\quad \left. + \frac{3.438 \cdot 10^{-10} \, [\mathrm{m^6\,mol^{-2}}]}{\bar{v}^3} \right) d\bar{v} \\
&= \bar{R}T \left(\ln \frac{\bar{v}_2}{\bar{v}_1} - 1.438 \cdot 10^{-5} \, [\mathrm{m^3\,mol^{-1}}] \left(\frac{1}{\bar{v}_2} - \frac{1}{\bar{v}_1} \right) \right. \\
&\quad \left. - 3.438 \cdot 10^{-10} \, [\mathrm{m^6\,mol^{-2}}] \left(\frac{2}{\bar{v}_2^2} - \frac{2}{\bar{v}_1^2} \right) \right).
\end{aligned}
\tag{8.22}
$$

Equation (8.22) gives the work from compression of hydrogen as a nonideal gas and as a function of the molar volume. This work is obviously negative since it is the input work done on the gas. Equation (8.21) gives the pressure as a function of molar volume, and thus the molar work versus pressure can be obtained. The absolute value of Eq. (8.22) divided by the molar mass of hydrogen is plotted as a function of pressure in Fig. 8.9, the gray line. Also, the ideal gas specific compression work is plotted, dashed black line. We can see that the ideal gas law is suitable up to 100 bar and that, at higher pressures, the viral equation predicts a higher work demand. At 350 bar and

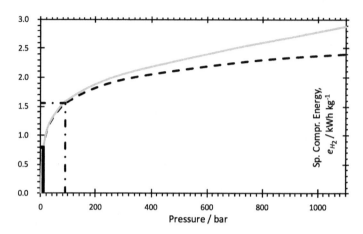

Figure 8.9 Specific work needed for compressing hydrogen from 1 bar, calculated using the viral equation with two terms (gray) and the ideal gas law (black dashed). Energy saved by starting the compression at 10 and 90 bar compression is also indicated.

700 bar, this deviation is 6 and 11% more than predicted by the ideal gas law. The viral equation of state is an approximation, also with limitation in precision. Considering isothermal compression is another approximation. Also, these calculations do not account for other irreversiblilites. Allover, the presented analysis therefore only gives a brief introduction to hydrogen compression and the related work. Regardless of the approximations, the analysis is useful as a starting point. Evaluating the work needed for compressing hydrogen, we should compare it to the specific energy content of hydrogen, 33 kWh kg^{-1}. In this context, around 10% of the work available is spent on compression hydrogen up to 700 bar.

8.2.2.3 Cryogenic Hydrogen

When transporting energy, both volumetric and specific energy should be enlarged if possible. This is the reason why liquid natural gas (LNG) process plants were developed. The same is in the case of hydrogen. For instance, it is by many envisioned that hydrogen can be produced in northern Europe, liquefied, and shipped by boat to Japan via the North East passage. The most dense form of technically feasible hydrogen is liquefied hydrogen. As can be seen in Fig. 8.8, this requires cooling to 20 K at 1 atm or 33 K at 13 atm.

Liquefaction is technically possible and requires cooling. Before liquefaction takes place, the hydrogen must be converted to para-hydrogen. This

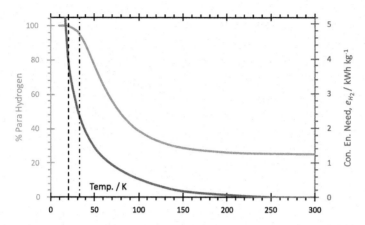

Figure 8.10 Content of parahydrogen at equilibrium at different temperatures (blue (light gray in print versions)) and the work required to convert normal hydrogen to para-hydrogen using a rankine refrigeration cycle (red (dark gray in print versions)) at different temperatures. The dashed lines indicate boiling points of hydrogen at 1 and 13 atm.

is an exothermic process, and heat must be removed at the relevant temperatures. Thus heat is taken at 20–33 K and delivered at room temperature. Considering a Carnot cycle for this otherwise reversible process, the coefficient of performance (COP) is

$$COP_{Carn.} = \frac{Q_C}{W} = \frac{Q_C}{Q_H - Q_C} = \frac{T_C}{T_H - T_C}. \tag{8.23}$$

Typically, a Rankine refrigeration cycle is used instead of a Carnot cycle, and the Rankine efficiency is typically in the order half of that of a Carnot (theoretical) cycle. One way to make cryogenic hydrogen is to cool the hydrogen in a heat exchanger using liquid nelium as coolant (nelium is a mixture btween heliumand neon).

Considering a Rankine cooling cycle and a heat exchanger, we can now evaluate how much work is needed to remove the conversion enthalpy of a kg of normal hydrogen into a kg of para-hydrogen. This is done using the tabulated data in Appendix C and dividing this value by the COP factor for the Rankine cycle. These data are plotted in Fig. 8.10 (red (dark gray in print versions) line). Two things are worth noticing in this graph: i) liquefaction at 13 atm (33 K) instead of at 1 atm (20 K) saves about half of the energy (2.4 versus 4.1 $kWh\,kg_{H_2}^{-1}$), and ii) the energy needed is

10–15% of the theoretical work needed to produce hydrogen from water ($33 \, \mathrm{kWh \, kg_{H_2}^{-1}}$).

In this example, we have not considered evaporation enthalpy, and the refrigerator can have much lower efficiency than 50%. Also, the specific energy for compressing hydrogen from 1 to 10 atm is around $0.80 \, \mathrm{kWh \, kg_{H_2}^{-1}}$, as indicated with a black solid line in Fig. 8.9. Under the given circumstances, first compressing hydrogen and then cooling it to liquefaction saves energy and next the investment cost of specialized equipment for cryogenic processes at very low temperatures.

8.2.2.4 Metal Hydride

Metal hydrides were for long regarded as a very promising hydrogen storage technology in the transport sector because of the nature of its characteristics: high volumetric density, high energy efficiency, and low pressure. The draw back of this technology is that it is slow and fueling time is very long. To understand this, we must know more about the principle of metal hydrides.

A metal hydride is based on the principle that hydrogen dissolves in metals in a solid state. This happens as hydrogen molecules H_2 adsorb on a metal surface. This adsorption is a chemisorption, which means that hydrogen forms a chemical bond or covalent binding with the metal. Physisorption is a different adsorption process, where only van der Waalian or dipole bindings take place. Once adsorbed, hydrogen further splits into hydrogen atoms and moves singularly into the bulk of the metal where they remain absorbed. (The terminology of sorption suggests that adsorption happens on a surface and that absorption happens into a volume. Within the field of metal hydrides, physisorption and chemisorption are sometimes labeled adsorption and absorption, respectively.) The hydrogen thus acts as ping pong (table tennis) balls between footballs, where footballs are a metaphor of the metal. Thus a metal hydride is formed. When hydrogen adsorbs onto the metal, heat is released, just like heat is released when vapor condenses on a surface. When storing hydrogen, this heat must be removed, and heat transport is a slow process. In fact, this is the rate determining step of metal hydrides as a technology [59].

The metal hydride is an equilibrium process, meaning that when heated to a given temperature, hydrogen is released at a certain pressure. If a metal hydride is at a given temperature, then one must send hydrogen above the equilibrium pressure in order for the metal to adsorb hydrogen. This adsorption, as already stated, is exothermic, and the material will increase

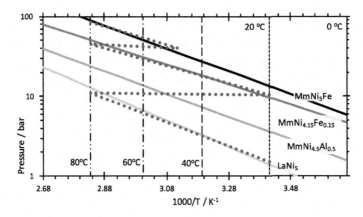

Figure 8.11 Pressure temperature equilibrium lines for different materials as functions of the inverse temperature. One possible three-step thermal swing adsorption process route is indicated (red dotted).

in temperature. Therefore adsorption requires cooling. For hydrogen to adsorb quickly, a large surface area is also needed. Therefore, the material is porous, that is, it consists of tiny particles of order 1–10 micrometers, and such materials are generally poor thermal conductors [60].

Metal hydrides can consist of different types of metals and usually as an alloy. Complex metal structures give more vacancies for hydrogen and also allow for tailoring adsorption properties. In Fig. 8.11, the pressure temperature equilibrium lines of four different metal hydrides are plotted as functions of the inverse temperature. These metal alloys are $MnNi_5$ (black), $MmNi_{4.15}Fe_{0.15}$ (dark gray), $MmNi_{4.5}Al_{0.5}$ (gray), and $LaNi_5$ (light gray). Mm denotes what is referred to as a mischmetal, typically (in wt.%) of roughly 48–50 Ce, 32–34 La, 13–14 Nd, 4–5 Pr and 1.5 other rare earths [61].

Despite the disadvantage of filling time due to heat management, metal hydrides have a very interesting use because of their strong temperature pressure dependency and their relation to heat [62]. This allows for using low-grade waste heat to compress hydrogen. This process is called thermal swing adsorption (TSA). The principle of TSA consists of filling up a metal hydride while cooling the reactor bed. When saturated, the reactor is sealed at the inlet, heated, and opened at the outlet. The hydrogen then comes out at a higher pressure. Low-grade waste heat in this context is typically of order 80°C.

In Fig. 8.11, a three-step swing adsorption process is indicated with a red dotted line. The first reactor bed consists of $LaNi_5$ and can compress hydrogen from 1.5 bar while cooling at 20°C up to 11 bar while heating to 80°C. The second reactor can contain $MmNi_{4.15}Fe_{0.15}$ and take hydrogen from 10 to 45 bar via cooling and heating at the same temperatures, and the third reactor can use a bed of $MnNi_5$ and can likewise take hydrogen from 40 to almost 90 bar pressure. Looking at Fig. 8.9, it is indicated (dash dot) how much work is saved per kg hydrogen by this TSA compressor, around 2/3rd of the total compression energy. Because a lot of waste heat is available in this temperature range, TSA is currently one of the most promising markets for metal hydrides.

8.2.2.5 Metal Organic Framework

Metal organic frameworks (MOF) consists of small organic molecules, typically aromates (C_6-rings) that are connected by metal clusters as a three-dimensional structure. Hydrogen can adsorb on these structures in very large and extremely dense amounts. One way to understand this is that while hydrogen forms a framework between metals in a metal framework, it is now the metal organic framework that creates large cavities where hydrogen is adsorbed and filled in the large holes [63].

While metal hydrides operate around and above room temperatures, MOFs operate around the boiling point of liquid nitrogen [64,65]. This means that metal hydrides cannot be used for TSA compression. Because of the need for cooling, long-term storage in a car does not fit as a market case either. Long-distance transportation as a more energy-efficient approach than liquid hydrogen is however an interesting market for this technology. The drawback is that the adsorption requires heat transfer, which is very slow in porous media like MOF and metal hydrides. When transporting hydrogen over larger distances, more volume also leads to more heat needed for offloading and more cooling for uploading the hydrogen. In competition with liquid hydrogen, MOF is shortcoming from a practical point of view, although it is more energy efficient.

8.2.2.6 Cavern and Grid Storage

Power-to-gas storage would have a very interesting potential in the future if today's natural gas network were replaced with a hydrogen pipeline. Then, ultrapure hydrogen would be available for a potential fleet of fuel cell cars. Also, the pipeline could function as a long-term energy buffer. As pointed out in Example 8.1, hydrogen has some potential for energy buffering in this manner; however, a very large energy network is needed for this.

Another example of power-to-gas storage that is envisioned for the future is to replace a few forthcoming air compression energy storage reservoirs with hydrogen storage. As indicated in Table 8.1, replacing compressed air energy storage with compressed hydrogen energy storage would increase the capacity by more than one order of magnitude. This is of course very promising; however, such energy storage facilities are uncommon, and even one extra order of magnitude is very little when comparing to a big tank of liquid hydrogen.

8.2.2.7 Carbon as a Hydrogen Carrier

Hydrogen can facilitate light long-range transportation for cars. However, for trucks, the volumetric energy density is too low. Hydrogen tanks would simply be too large for this type of heavy-duty long-range vehicles. Looking at Fig. 8.2 (p. 150), we can see that liquid fuels offer better volumetric energy density and that hydrocarbon-chained fuels, like diesel and jet fuel, are the best. Both liquid biogas, biodiesel and biojet fuel, offer high volumetric density that is renewable. These energy storage media can be upgraded using hydrogen. One example is to mix hydrogen into biogas, like it is in power-to-gas, and then distribute it as renewable hydrogen-enriched biogas or natural gas.

Biofuels are about energy production rather than energy storage per se. Biofuels just happen to be storable as they are. When first looking at biofuels storage and distribution aspects, it is worth mentioning some current trends within biogas (BG). A very interesting trend within BG is that it is currently and foremost produced locally from small and medium units with agreements on delivering the produced gas to a local bus fleet or other public services. The change in this trend is that liquefied biogas (LBG) is now also produced in larger plants because LBG can be sold over a larger area since it makes sense to distribute LBG ten times longer than BG. Also, this opens a different business model as the producer has a larger market than a few local costumers. In other words, the trend is towards larger biogas plants that produce LBG for long-range distribution and large markets.

8.3 REUSE OF HYDROGEN: FUEL CELLS

Electrochemical hydrogen production has business cases for markets where ultrapure hydrogen is needed or surplus electric energy take-up is needed.

Ultrapure hydrogen is needed for niche markets (e.g., electric generator cooling or food industry) and for low-temperature fuel cells. Currently, low-temperature fuel cells is an emerging technology, albeit with an exponential growth. Several car manufacturers produced fuel cell vehicles, and the amount of fueling stations have started to form continuous networks in several countries and regions.

In the context of this chapter and book, our main focus is on fuel cells with hydrogen from electrolysis. Currently, this hydrogen does not represent the main source of hydrogen in the market, and several types of fuel cells operate on other types of hydrogen or fuels. Moreover, biogas and liquid biofuels are energy sources that come in a storable format, and they could be reused in fuel cells as well. Although hydrogen from electrolysis is the main focus here, other relevant fuel cell technologies are described briefly to give the reader a first introduction to different fuel cell technologies.

8.3.1 Fuel Cell Thermodynamics

Generally, the thermodynamics of a fuel cell reaction is very similar to that of water electrolysis, with the exception of changing sign of values and directions of every reaction. In this respect, the thermodynamics is already given in Section 8.1.1. The fuel cell reaction becomes Eq. (8.24), and the performance Eq. (8.25), where g/l denotes vapor or liquid water, and η_i sum up all nonohmic potential losses:

$$H_{2,(g)} + \tfrac{1}{2}O_{2,(g)} \Rightarrow H_2O_{g/l}, \tag{8.24}$$
$$E^{cell} = E^{rev} - rj - \eta_i. \tag{8.25}$$

When operating fuel cells, a few additional aspects come into consideration; however, these typically have to do with carbon. One challenge is in relation to the formation of carbonates in the electrolyte if the pH is at too high level, and the other to the anode side in low-temperature fuel cells, the electrode where the fuel is oxidized when fueling with hydrocarbon.

One carbon-related difficulty in fuel cell catalysis is the formation of solid carbonates. This stems from the equilibrium reaction between carbon dioxide, hydroxide ions, and cations. Equation (8.26) exemplifies this with an arbitrary metal divalent ion Me^{2+}, forming a metal carbonate in a highly alkaline solution (high pH). When the pH is high (e.g., > 10), the equilibrium reaction in Eq. (8.26) shifts to the right. This means that, for alkaline

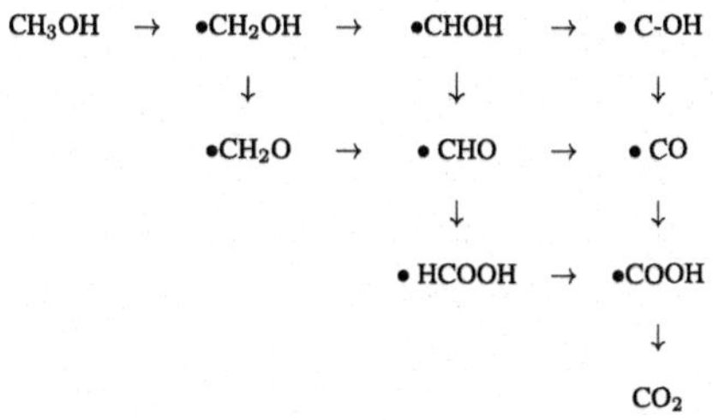

$$CH_3OH \rightarrow \bullet CH_2OH \rightarrow \bullet CHOH \rightarrow \bullet C\text{-}OH$$
$$\downarrow \qquad\qquad \downarrow \qquad\qquad \downarrow$$
$$\bullet CH_2O \rightarrow \bullet CHO \rightarrow \bullet CO$$
$$\downarrow \qquad\qquad \downarrow$$
$$\bullet HCOOH \rightarrow \bullet COOH$$
$$\downarrow$$
$$CO_2$$

Figure 8.12 Methanol reaction pathways. The bullet indicates a chemical bonding between the partially oxidized compound and the catalyst [66].

fuel cells, air cannot be used unless it is absolutely free of carbon dioxide.

$$CO_{2,(aq.)} + 2OH^- + Me^{2+}_{(aq.)} \leftrightharpoons MeCO_{3,(s)} + H_2O_{(l)}. \tag{8.26}$$

Carbon monoxide CO poisoning is a phenomenon that takes place when carbon monoxide adsorbs onto the platinum catalyst particle so strongly that further reactions are inhibited and other reactions on the catalyst blocked. Impurities in the hydrogen gas, for instance, from reforming, are present in the form of CO, and platinum catalysts are inhibited, commonly referred to as poisoned. The other way that carbon monoxide poisoning can occur is when hydrocarbon fuels are used [66]. Liquid fuels are more energy dense and thus a good solution in many instances. One example is methanol, H_3COH, which is liquid at room temperature and pressure. During the oxidation, several steps take place, and CO poisoning is inevitable on platinum, as can be seen form the reaction patterns in Fig. 8.12. Generally, when oxidizing organic fuels in a fuel cell, a reaction step via carbon monoxide is almost inevitable. Two strategies to prevent carbon monoxide poisoning are advanced catalysis (often adding ruthenium to the platinum) or raising the temperature since carbon monoxide desorbs from platinum above a temperature around 140–150°C.

8.3.2 Fuel cell technologies

Fuel cells as a concept have been around for more than a century. As a system for energy conversion, it is often attributed to Sir W. Groove, whereas the principles of it are attributed to C. F. Schönbein who reported the principles in 1839 [66–68]. Before this, in 1802, Davy reported similar phenomena [69], though it is debated to what extent this report represents a hydrogen fuel cell.

Early adaptation of fuel cells for power came more than a hundred years later, however, in the USA space programs, now more than half a century ago. Since then, a lot of development has been made, and several types of fuel cells have emerged, each tailored for different types of fuels and purposes. Based on the current market impact and maturity, descriptions of the different technologies are given in the following.

Before outlining the different fuel cell technologies, it should be emphasized that fuel cells are energy-converting units rather than an energy source, similarly to combustion units like ICEs and turbine machines. Moreover, the context of this chapter is hydrogen as an energy storage medium, and therefore the main focus is on converting hydrogen into electricity. However, several fuel cells are capable of converting other fuels as well as hydrogen, like, for example, natural gas, ethanol, and other hydrocarbons. Some fuel cells can even use coal if appropriately reformed into carbon monoxide (Eq. (8.19)) or hydrogen from a water gas shift reaction (Eq. (8.20)); however, the main focus in this book will be kept on hydrogen when it comes to fuel cell description.

8.3.2.1 Proton Exchange Membrane Fuel Cell, PEMFC

In the context of energy storage and hydrogen, there is one fuel cell technology that is by far more developed than the others, and this is the PEMFC. PEM is an acronym for either polymer electrolyte membrane or proton exchange membrane, whereas FC is for fuel cell. Here, PEM is used for proton exchange membrane since protons are exchanged in the membrane and there are polymeric membranes that can be used for all kinds of electrolytes (e.g., in alkaline systems and lithium ion batteries).

The core of the PEMFC is the membrane electrode assembly (MEA) [70]. This consists of a membrane coated with a catalyst layer (CL) and is therefore also commonly referred to as catalyst coated membrane (CCM). In Fig. 8.13, this is indicated yellow (light gray in print versions), where light yellow (superlight gray in print versions) is the membrane, and darker yellow (light gray in print versions) is the electrode or catalyst layer.

The membrane must be insulating with respect to electrons, conductive with respect to ions, and separate the reacting gases from each other in order for its function to sustain. The membrane material most widely applied is Nafion, which is a polymer similar to polytertafluorethylene (PTFE or Teflon). The polymer is in turn equipped with functional groups that hold sulphonic active groups ($-SO_3^-$) attracting protons and water. When fully humidified by liquid water, there can be as much as 22 water molecules

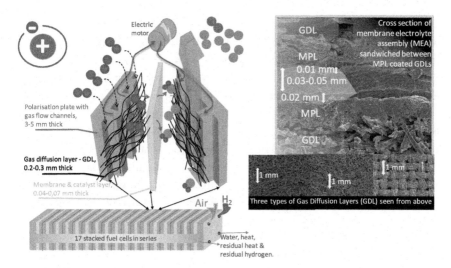

Figure 8.13 A proton exchange membrane fuel cell, PEMFC. The unit cell consists of the membrane electrolyte assembly (MEA), which is the catalyst layer (CL) on each side of the membrane. The MEA is a sandwich between microporous layers (MPL), which are on top of the gas diffusion layers (GDL), and polarization plates with gas flow channels.

per sulphonic group, whereas it is usually about half when humidified by vapor. The ionic conductivity is of order 10 S/m for normal operational conditions [48]. More water gives better conductivity and vice versa. In a modern fuel cell, the membrane thickness ranges from 20 to 60 micrometers. A Nafion typically comes in a thickness of 1, 2, 5, or more in mills (milliinches or 25.4 µm) and expands around 15% when submerged in hot liquid water [71]. Several new membranes are introduced on the market, and they have similar properties as Nafion, albeit with different durability and cost.

The electrodes must have electric and ionic conductivity in addition to being permeable for the reacting gases, durable, cheap, and leading the water product away [72]. This is a demanding set of requirements that comes on top of perhaps the most demanding request: low Tafel or equivalently Butler–Volmer overpotential. With respect to the Tafel overpotential, this is not a significant issue for the anode, where the hydrogen reaction takes place; however, it is a major drawback for the oxygen reduction reaction at the anode. The cathode Tafel overpotential is almost 0.2 V at the very smallest usable current densities of a PEMFC, meaning that around 17 % of the potential work is lost as heat due to the friction of transferring electrons between oxygen and the electrode. The solution to minimize this lost work is usage of platinum, something that for a very long time was the bottleneck for commercial and affordable PEMFC development. During the last 5–10 years, this issue is overcome. Moreover, the way that the catalyst layer can have all the requested properties is by a gas electrode. In this electrode, very small carbon particles (of order 100 nanometers) are mixed together with small pieces of the membrane material to form a porous structure. The carbon particles are in turn covered with even smaller nanoplatinum particles. The key for a good catalyst is being in contact with both membrane material and gas. Since this requires three different substances to be in contact (platinum/catalyst, dispersed material, and reacting gas), *three phase boundary* (TPB) lines are formed. Having a large enough amount of TPB is one way to recognize a good catalyst. As said, the carbon particles carrying the catalyst are very small, and using a regular scanning microscope (SEM) will only display the thickness of such a layer. In Fig. 8.13. the catalyst layer photographed by using a SEM is indicated with dark transparent yellow (light gray in print versions), with the anode and cathode being of order of 10- and 20-μm thick, respectively.

Whereas the MEA is the core of the PEMFC, there is a series of materials that support it. From the outside this starts with gas feed channels. Next, it is a fibrous layer where the gas is distributed more uniformly across the MEA and an even finer distribution layer is added. The layers are commonly referred to as gas diffusion layers (GDL) and microporous layers (MPL), where GDL is the fibrous layer, and MPL is the finer distribution layer. The gas distribution channels, GDL and MPL, both deserve a more detail explanation, starting from the MEA, going towards the bipolar plate [73]. Adjacent to the CLs, MPLs are placed. The role of the MPL is to distribute gas and electrons and to remove water. It is a lot of controversy to what the MPL exactly does and the mechanisms behind it; however, there is a strong agreement that it significantly improves the performances of the PEMFC. The MPL consists of small (similar to the CL) carbon particles and PTFE and is commonly made by having a carbon and PTFE dispersion sprayed or smeered on top of a GDL [60]. Because of this manufacturing procedure, the MPL exists partly as a single layer and partly as a composite with the GDL.

The GDL is on the outside of the MPL. A GDL consists of carbon fibers put together in a layered form. This layer typically has a porosity of around 80% and a thickness varying between 100 and 400 μm [74] and is intended to remove water and for the feed gases to flow through. The name gas diffusion layer is misleading because diffusion is not the most important transport mechanism and because liquid water is common during operation. Nevertheless, GDL is the common term. There are many different types of GDLs, and some of them are shown in Fig. 8.13.

The gas is distributed in the fuel cell on a macrolevel by the gas distribution channels that lay on top of the GDL. This occurs on each side of the GDLs, and in Fig. 8.13, this is illustrated with so-called parallel flow channels. The feed channels can have other configuration as well, for example, serpentine like and spiral like; however, parallel flow ones are often selected for simplicity.

Moreover, the gas feed channels are a part of the polarization plates, more commonly referred to as the bipolar plates. On such plates, the hydrogen fuel stream on the one side and the air on the other side. Thus the bipolar plate functions as a connector between fuel cells stacked up in a series configuration.

As is illustrated quite clearly in Example 8.4, substantial heat is produced in addition to electric power. This heat must be removed to avoid overheating and in turn severe degradation of the fuel cell. Two main strategies appear for this, where one is air cooling, and the other is internal water/oil cooling channels [72,75]. Whereas the first strategy consists of purging sufficient air through the cathode side of the PEMFC, the second strategy requires liquid cooling channels inside the bipolar plates. Such channels can be realized by assembling two plates together to form a single bipolar plate with cooling channels inside.

In fact, the large portion of heat from fuel cells is by many considered valuable for residential heating through combined heat and power (CHP) [76]. In a CHP system, the heat is a biproduct, meaning that a fuel cell is designed for electric energy supply rather than thermal.

Example 8.4: A First-Order PEMFC Heat and Work Model.

Set up a model for a PEMFC potential and power density as a function of current density (in A/cm^2). Plot the model graphically and indicate thermal and electrical energy and power.

a) Use the thermodynamic data for a PEMFC at 1 bar and room temperature given by Table 8.2, a conductivity of 10 S/m [48] for a Nafion 112, and Tafel kinetics of $\eta_T = 0.45 + 0.13 \log j$ with current density in A/cm^2 [77] without accounting for mass transport limitations.

b) Add the concentration over potential losses using the equation

$$\eta_{Cons} = \frac{\bar{R}T}{4F} \ln \left[1 - \frac{\bar{R}T}{4Fh_{O_2,eff}\, p_{O_2,b}} j \right],$$

where $h_{O_2,eff}$ is the effective mass transport coefficient, and set it to 0.16 W/cm^2 bar (equivalent to 1600 m/s). What happens if the effective mass transfer coefficient lowers by a factor of 2?

c) Derive the expression for the concentration overpotential based on Fick's law of diffusion and the Nernst equation for mass flux and current density.

Solution:

a) *First, we set up the expression for the cell potential:*

$$E^{cell} = E^{rev} - rj - \eta_T,$$

where the reversible potential is given by

$$E^{rev} = E^o - \frac{\bar{R}T}{4F} \ln \frac{p^o_{O_2}}{p_{O_2,b}} = 1.23 \text{ [V]} - \frac{8.314 \cdot 298}{4 \cdot 96485} \ln \frac{1}{0.2} \text{ [V]} = 1.22 \text{ [V]}.$$

The specific resistance of the cell becomes

$$r = \frac{\delta}{\kappa} = \frac{0.00063 \text{ [cm]}}{0.1 \text{ [S/cm]}} = 0.063 \text{ [}\Omega \text{ cm}^2\text{]}$$

when considering a 2-mill-thick membrane swelled from water absorption. The potential and power density thus become

$$E^{cell} = 1.22 - 0.063j - 0.45 - 0.13 \log j \text{ [V]}$$

and

$$P = 1.22j - 0.063j^2 - 0.45j - 0.13j \log j.$$

In addition to the heat irreversibly dissipated from the ohmic friction and the friction of electron transfer at the electrodes, reversible heat from the reaction entropy is also released. This is the difference between the thermoneutral potential and the reversible potential.

The energies from the fuel cell are plotted in Fig. 8.14, where the reversible heat is indicated with checkerboard pattern, the irreversible heat from the ohmic friction is indicated by vertical pattern, and the Tafel heat from friction of electron transfer by horizontal pattern. The cell potential at this point is indicated by a dotted line.

b) *Accounting also for the concentration overpotential, we obtain*

$$E^{cell}/V = 1.22 - 0.063j - 0.45 - 0.13\log j - 0.0064\ln[1 - 0.20j]$$

units in V and

$$P = 1.22 - 0.063j^2 - 0.45j - 0.13j\log j - 0.0064\ln[1 - 0.20j]j$$

with units in Wcm^{-2}.
This behavior is plotted with the thick black line in Fig. 8.14. We can see that once the concentration polarization effects start to take place, the cell potential and power drop very rapidly. This is however not entirely a true behavior since the flow rate of air is increased with the current rate and the effective mass transport coefficient increases with the current density. The dash-dot line indicates a mass transport coefficient reduced by a factor of two and shows that the mass transport limitation heavily depends on the air feed. Regardless of its simplicity, this simple model shows how the main transport processes of charge and mass take place and how they can be modeled as a first-order simplification.

c) *Fick's law of diffusion gives the molar flux as*

$$J_{O_2} = -D_{O_2}\frac{dC}{dx}.$$

The Nernst equation gives the relation between the mass flux and current as $J_{O_2} = -\frac{j}{4F}$. By insertion into each other and next introducing the ideal gas law ($C_i = p_i/\bar{R}T$), we obtain

$$\frac{j}{4F} = D_{O_2}\frac{\Delta C}{\Delta x} = \frac{D_{O_2}}{\bar{R}T}\frac{\Delta p_{O_2}}{\Delta x} \Leftrightarrow p_{O_2,b} - p_{O_2,s} = \frac{\bar{R}T}{4FD}\Delta x\,j. \quad (8.27)$$

The concentration overpotential is in turn given as

$$\eta_{Cons} = -\frac{\bar{R}T}{4F}\ln\frac{p_{O_2,b}}{p_{O_2,s}}, \quad (8.28)$$

where $p_{O_2,b}$ and $p_{O_2,s}$ refer to the partial pressure of oxygen in the bulk phase (air) and at the surface where the reaction takes place. Equation (8.28) requires the ratio between the partial pressures rather than the difference between them, so that Eq. (8.27) must be rearranged into Eq. (8.29) before inserted into Eq. (8.28):

$$\frac{p_{O_2,b}}{p_{O_2,s}} = 1 - \frac{\bar{R}T}{4F\bar{h}_{O_2,eff}\,p_{O_2,b}}j, \quad (8.29)$$

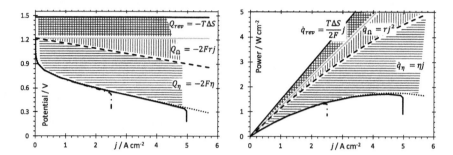

Figure 8.14 Left: Potential energy and corresponding thermal energy sources as functions of current density for a PEMFC producing liquid water. Right: Power density and heat density for the same PEMFC conditions.

where $\bar{h}_{O_2,eff}$ replaces the diffusion coefficient D_{O_2} divided by the layer thickness Δx as an empirical fit similar to what the Newtonian heat transport coefficient does for thermal conductivity and film thickness layer. Thus we obtain the following expression for the concentration overpotential:

$$\eta_{Cons} = -\frac{\bar{R}T}{4F} \ln\left[1 - \frac{\bar{R}T}{4F\bar{h}_{O_2,eff}P_{O_2,b}}j\right]. \tag{8.30}$$

Yet a remark on the simplifications of this model, is that the bulk partial pressure of oxygen $p_{O_2,b}$ is rarely 0.2 bar and for at least two reasons: If considering dry gases, the partial pressure would be close to 0.2 bar, but due to the pressure drop in the flow channels, the pressure is much higher (factor of 2–5). The other reason is that once above $46°C$, the partial pressure of water starts to increase beyond 0.10 bar. At 70 and $80°C$, the partial pressure of water is 0.31 and 0.47 bar, respectively. At 1 bar total pressure, the partial pressure is thus reduced to around 0.1 bar at $80°C$ with saturated air. The point is that much more complex and realistic models can be made from the simple model given here.

The PEMFC has been around for decades and with high promises given, promises beyond reach when first given. Today, department of oil and energy (DOE) in the USA can be seen as the most reliable source of information to what can be expected. The DOE sets targets for what performance to be met at what time in order for the PEMFC to be an established and cheap power source by 2030. The assessment constitutes of most needed standards from MEA laboratory measurement protocols to expected price and platinum content of all components of a PEMFC. The targets, well within reach, are that an entire fuel cell system (tanks, controllers, motor, etc.) shall [36]:

- transport a car for 600 km and weight 90 kg in 2020 and 65–70 kg in 2030;
- be fueled (full fueling cycle) in 3.3 minutes by 2020 and in 2.5 minutes in 2030;
- cost around 4 000 US$ in 2020 and around 3 000 US$ in 2030.

Currently, a fuel cell system costs around 30 000 US$; however, this price is not for mass produced fuel cell systems. The price is estimated to drop from 30 to 6 thousand dollars just by the means of mass production [36]. For reference, a small car engine (ICE with no transmission) with around 100 hp weighs 80–100 kg. Adding transmission and a full tank (50 L gasoline) easily ramps today's car propulsion system up to 150 kg, meaning that a fuel cell system will weigh less than half of an ICE system when fully matured.

8.3.2.2 Direct Methanol Fuel Cell, DMFC

The direct methanol fuel cell (DMFC) has many similarities with the hydrogen PEMFC. The two main differences are that on the anode side, the hydrogen gas is replaced with a liquid solution of water and methanol and that the catalyst is a combination of platinum and ruthenium [78]. The methanol reaction path on the catalyst can be very complex and consists of different routes, as indicated in Fig. 8.12. The reason for introducing ruthenium is to prevent carbon monoxide poisoning of the catalyst as this occurs to some extent and the carbon monoxide otherwise blocks the active catalyst sites. DMFC has its advantages in being a steady and reliable fuel cell although it offers less power than other fuel cells.

8.3.2.3 Solid Oxide Fuel Cell, SOFC

The solid oxide fuel cell (SOFC) is very similar to the SOEC, except that the current flows the other way and it consumes oxygen and hydrogen [79]. A great advantage of the SOFC is that it can run on almost any hydrocarbon fuel in addition to pure hydrogen. This can be realized in two ways; either by prereforming the hydrocarbon, for example, natural gas, or by internal reforming. The reforming (internal or externally prereformed) is a highly endothermic reaction and requires heat. The SOFC generates a lot of extra heat from the ohmic electrode overpotential and the reversible (entropic) heat. This heat can be used not only for the reforming, but since the heat is of high quality (temperature), the SOFC surplus heat is useful for several other applications. Thus the SOFC is considered a very promising candidate for combined heat and power.

8.3.2.4 Alkaline Fuel Cells, AFC

Alkaline fuel cells (AFC) have been around for a very long time. In fact, early usage was in the Apollo space program, where water and electricity was supplied from an alkaline fuel that delivered 1.5 kW and weighed 113 kg [80], one percentage of today's PEMFC power density. The advantage of this technology is similar to the alkaline electrolysis in that one can use nickel and nickel oxide electrodes and that these are durable and cheaper compared to platinum. The limitation of the AFC is that if carbon dioxide gets into the electrolyte, then one will form alkaline carbonates that solidify in the electrolyte. This is very detrimental in an AFC because this fuel cell traditionally has had a circulating electrolyte. This means that the AFC consists of an open space between the electrodes, where the electrolyte flows through, and two gas phases that are outside the two electrodes. To prevent the carbonate formation, the AFC is typically operated using pure oxygen and pure hydrogen. For this reason (pure oxygen), the AFC is typically suited for stationary power supplies.

8.3.2.5 Other Fuel Cell Technologies

The phosphoric acid fuel cell (PAFC) uses a solid acid carrier that carries the acid, and they together constitute a membrane [81]. The carrier is typically based on silicon carbide particles (SiC), and systems have been reported to operate for in the order of 40 000 h without significant electrolyte losses [82]. The PAFC typically operates around 200°C and function in most ways very similarly to a PEMFC. The main advantage of the PAFC is that it is not sensitive to carbon monoxide poisoning. The draw back is its low power density and that is has been a challenge to have the acid carrier keep the acid for long enough. The polybenzimidazole (PBI) fuel cell can be seen as a modernized version of the PAFC since it uses a polymer-based electrolyte carrier [83]. When operated well above 140°C, difficulties with carbon monoxide catalyst poisoning go away. Both the PAFC and PBI fuel cells have fairly good kinetics. Because of high operation temperatures (140–220°C), this technology is best suited for stationary use and possible also in combination with heating, for example, radiator systems in residents.

The molten carbonate fuel cell (MCFC) is the same as the MCEC, except operating with the current in the opposite direction [84]. Oxygen from the cathode side is reduced and dissolved along with carbon dioxide as the carbonate ion CO_3^{2-}. Although carbon dioxide goes into the cathode description, it does not have to if then oxygen ions only "jump" between the carbonate ions. The advantage of the MCFC is that it operates at so high temperatures (500°C and above) that it has a great fuel flexibility. Again, above 300–400°C, catalysis for different fuels is convenient and cheap. Also, because of the high operation temperature, the MCFC offers high quality heat. The need to be at high temperature makes need for stationary power and combined heat and power a good market segment for this technology.

8.3.2.6 Fuel Cell Technology Overview

Several fuel cell technologies are available, and only seven of them are described here. Fuel cells are promising electrochemical energy converters both for power and combined heat and power. In the context of this book, fuel cells are described mainly with respect to energy stored in the form of pure hydrogen. All fuel cells are adequate for this, but with the exception of the PEMFC and partly AFC, they mostly have their niches in converting hydrocarbons from various sources. One can argue that biofuels are stored energy; however, bioenergy is about energy collection, reaction, and processing rather than energy storage. Regardless of relevance to energy storage and the hydrogen energy context, a brief summary of different fuel cell technologies can be justified here. Table 8.4 lists the technologies with their characteristic electrolyte charge transfer, characteristic cathode reaction, and potential niche anode reactions.

To sum up this chapter; hydrogen is currently mainly produced to meet a need for hydrogen and therefore from fossil fuels. The emerging market for hydrogen production by electrolysis stems from the need of handling the surplus of electric energy from intermittent renewable electric energy sources. The technologies that appear to dominate electrolysis are alkaline water electrolysis, proton exchange membrane electrolysis, and solid oxide electrolysis. Moreover, hydrogen storage appears to be realized through two main technologies, compressed hydrogen for automotive industry and liquid hydrogen for long-distance transportation. Other storage technologies can meet niche markets for long-term storage and TSA compressors. In reusing hydrogen, the PEMFC is currently the most robust technology for large-scale mass production, expected to take place from around 2020. SOFC is also likely to be commercialized on a large scale because of its fuel flexibility and benefits within combined heat and power.

Table 8.4 Overview of different fuel cell technologies, electrode reactions, electrolyte charge carriers, and temperature operating range

Fuel cell technology	Anode reaction	Electrolyte charge carrier	Cathode reaction	Temp /°C
PEMFC	$H_2 \rightarrow 2H^+ + 2e^-$	H^+	$\frac{1}{2}O_{2,air} + 2H^+ + 2e^- \rightarrow H_2O$	-20–120
DMFC	$CH_3OH + H_2O \rightarrow CO_2 + 6H^+ + 6e^-$	H^+	$\frac{3}{2}O_{2,air} + 6H^+ + 6e^- \rightarrow 3H_2O$	0–80
SOFC	$CH_4 + 4O^{2-} \rightarrow CO_2 + H_2O + 8e^-$	O^{2-}	$2O_{2,air} + 8e^- \rightarrow 4O^{2-}$	600–900
AFC	$H_2 + 2OH^- \rightarrow 2H_2O + 2e^-$	H^+	$\frac{1}{2}O_2 + H_2O + 2e^- \rightarrow 2OH^-$	0–80
PAFC	$H_2 \rightarrow 2H^+ + 2e^-$	H^+	$\frac{1}{2}O_{2,air} + 2H^+ + 2e^- \rightarrow H_2O$	180–220
PBI	$H_2 \rightarrow 2H^+ + 2e^-$	H^+	$\frac{1}{2}O_{2,air} + 2H^+ + 2e^- \rightarrow H_2O$	140–180
MCFC	$H_2 + CO_3^{2-} \rightarrow H_2O + CO_2 + 2e^-$	CO_3^{2-}	$\frac{1}{2}O_{2,air} + CO_2 + 2e^- \rightarrow CO_3^{2-}$	500–800

PROBLEMS

Problem 8.1. Hydrogen and Temperature.

a) Using the thermodynamic data given in Table 8.2, plot the standard potential for hydrogen oxygen reaction between 25°C and 900°C.

b) Consider making hydrogen with 0.2 V total overpotential (ohmic and Tafel). Plot the electric energy efficiency relative to 25°C. Is this a realistic comparison given the information in this chapter?

c) Look at Fig. 8.7. State-of-the-art technology has membrane thicknesses that lower the ohmic resistance of the cell by a factor 5, as the membrane becomes a tenth in thickness, and this constitutes about half of the total losses. Plot the new graph for the ceria scandia stabilized zirconia (CSSZ) and evaluate the energy needs and electric efficiency for the two designs.

Problem 8.2. Alkaline vs. PEM Water Electrolysis.

Compare the efficiency of alkaline water electrolysis and PEM water electrolysis at 7 $kA\,m^{-2}$ from Figs. 8.5 and 8.6.

a) What are the energy efficiencies?

b) What is the main contributor to the energy loss in the two technologies?

c) How will these efficiencies develop when the current doubles?

Problem 8.3. PEM Fuel Cell Heat Production.

For a PEM fuel cell, consider the liquid thermodynamic data in Table 8.2. The membrane is 35-μm thick and has a conductivity of 8.7 $S\,m^{-1}$. The anode overpotential is constant at 1 mV. On the cathode, the overpotential is given as $\eta_{katode} = -0.5 - 0.07\ln j$ (current density given as $A\,cm^{-2}$).

a) What is the reversible potential at 25 and 90°C?

b) Determine the cell potential and power when the current efficiencies are 0.1, 0.3, 0.5, and 0.8 $A\,cm^{-2}$?

c) Determine the efficiency and different contributions to loss in efficiency.

d) What if at 1 A/cm^2 oxygen would not reach the cathode due to diffusion limitations? Sketch this scenario graphically.

e) What is the heat production at 0.5 A/cm^{-2}, including the reversible heat?

f) As described in Fig. 8.13, the PEM fuel cell MEA is sandwiched between several porous transport layers. Consider this layer to be 300-μm thick and to have a thermal conductivity of 0.30 $W/K\,m$ [74]. How large will the temperature drop across this layer be?

SOLUTIONS

Solution to Problem 8.1. Hydrogen and Temperature.

a–b) We calculate the standard potential as $E^o(T) = -\frac{\Delta \bar{h}^o - T\Delta \bar{s}^o}{2F}$, using the liquid water properties below 100°C and the gas water properties above.

The efficiency is calculated as follows:

$$\varepsilon = \frac{E^o(T) + 0.2 \,[\text{V}]}{E^o(T = 298 \text{ K})}.$$

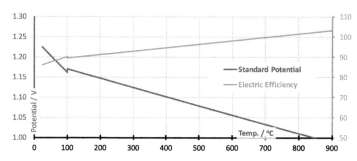

At very high temperatures, an overpotential of 0.2 V is actually very realistic, even at higher current densities. At lower temperatures, however, overpotentials of 0.4–0.6 V are much more realistic.

c) The overpotential is proportional to the change in resistance as there are hardly any electrode kinetic losses at this temperatures. The old (left) and the new (right) graphs thus look as follows:

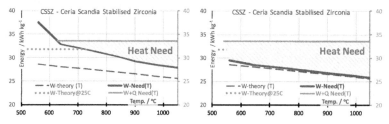

It can be seen that although the electricity need is very low, the heat requirement is larger.

Solution to Problem 8.2. Alkaline vs. PEM Water Electrolysis.

a) The energy efficiencies are calculated from the open cell potential divided by the cell potential. From graphical readings (approximately) the energy efficiencies are (1.22/2.6 =) 46% and (1.22/2.0 =) 61% for the alkaline and the PEM water electrolyzers, respectively.

b) The main contributor to the loss in energy efficiency is ohmic potential for the alkaline cell, whereas it is the anodic overpotential that is the largest contribution to the loss for the PEM water electrolyzer.

c) Because of the logarithmic behavior of the electrode overpotential, this potential will dominate and remain closely in size when doubling the current density for the PEM electrolyzer. For the alkaline water electrolyzer, the ohmic potential loss will double in size and lead to a stronger worsening of the energy efficiency for this electrolyzer when doubling the current.

Solution to Problem 8.3. PEM Fuel Cell Heat Production.

a) E^o is 1.22 V and 1.17 V at the two temperatures: $E^o = -\frac{\Delta \bar{h}^0 - T\Delta \bar{s}}{2F}$.

b) The cell potential is given as $E^{cell} = E^o - rj - \eta_{kat} + \eta_{anode}$.
   At 90°C, we obtain:

E^{cell}	0.940	0.874	0.824	0.739	0.695	0.650	0.626
$P/\text{kW m}^{-2}$	0.2	0.4	0.8	2.2	3.5	5.2	6.3
$j/\text{A cm}^{-2}$	0.02	0.05	0.1	0.3	0.5	0.8	1
rj/V	0.001	0.002	0.004	0.012	0.020	0.032	0.040
$\eta_{cath.}/V$	0.001	0.001	0.001	0.001	0.001	0.001	0.001
$\eta_{an.}/V$	−0.226	−0.290	−0.339	−0.416	−0.451	−0.484	−0.500

c)

E^{cell}	0.940	0.874	0.824	0.739	0.695	0.650	0.626
$\varepsilon/\%$	80	75	71	63	60	56	54
f_{rj}			0.01	0.03	0.04	0.06	0.07
$f_{\eta_{kat}}$			0.00	0.00	0.00	0.00	0.00
$f_{\eta_{anode}}$			0.99	0.97	0.96	0.94	0.92

d) Look at Example 8.4 and Fig. 8.14. The potential will drop abrupt.

e) Heat flux density due to irreversible loss in efficiency equals the power density divided by the efficiency minus the power density:

$$\dot{q}_{irrev.} = P\frac{(1-\varepsilon)}{\varepsilon} = 3.5\frac{0.4}{0.6} \ [\text{kW/m}^2] = 2.33 \ [\text{kW/m}^2].$$

The reversible heat stems from the entropy change in the fuel cell:

$$\dot{q}_{rev.} = -\frac{T\Delta\bar{s}^o}{2F}j = \frac{363 \ [\text{K}] \cdot 167 \ [\text{J/K mol}_{H_2}]}{2 \ [\text{ekv/mol}_{H_2}] \cdot 96485 \ [\text{C/ekv}]}500 \ [\text{A/m}]$$

$$= 1.57 \ [\text{kW/m}^2].$$

The total heat generation is thus 3.9 [kW/m²].

f) Fourier's first law gives.

$$\dot{q} = -k\frac{dT}{dx}.$$

The heat produced in the MEA will separate (more or less) equally to each side, so the heat flux through the given layer is 1.95 kW m^{-2}. Rearranging gives

$$\Delta T = \dot{q}\frac{\Delta x}{k} = 1950 \ [\text{W/m}] \frac{3 \cdot 10^{-4} \ [\text{m}]}{0.3 \ [\text{W/K m}]} = 2.0 \ [\text{K}].$$

For traditional membranes, the temperature should not exceed 95°C since the membrane degrades much quicker [85,86]. Doubling the current density would thus more than double the temperature gradient. Keeping track of your temperature profile is for this and many other reasons very important!

CHAPTER 9

Supercapacitors for Energy Storage and Conversion

Whenever electrochemical energy storage of mobile applications is discussed, the terms supercapacitor, ultracapacitor, double-layer capacitor, or electrochemical capacitor are likely to be brought up. These are synonyms for the term supercapacitor. The reason for considering supercapacitors is their extremely high power density in terms of both mass and volume. Albeit supercapacitors are electrochemical, they do not include red-ox reactions. At the same time they are sensitive to changes in concentration. Since there is no red-ox reactions and thus no friction for charge transfer, these systems can take enormous loads in very short time and with high efficiency. For the same reasons (i.e., no red-ox reactions), these systems store very little energy. This chapter explains the principles of conventional capacitors, supercapacitors, energy and power dimensioning, and some very interesting features of salinity gradient energy extraction.

9.1 CONVENTIONAL CAPACITORS

A conventional double plate capacitor consists of two parallel plates, typically of aluminum, separated by a thin polymer film. These capacitors are therefore often referred to as thin-film capacitors. The capacitance of a capacitor or capacitor system is defined as how much the charge can change for a given change in potential:

$$C = \frac{\Delta Q}{\Delta E}. \tag{9.1}$$

The polymer film thickness represents an internal distance d. When applying a current I for some time Δt or charging this pair of plates to a certain voltage E, one of the plates will experience a deficit of electrons, and the other a surplus relative to the natural being of these plates. This is illustrated (left) in Fig. 9.1. The corresponding charge is the product of the time and current applied:

$$Q = I\Delta t. \tag{9.2}$$

Engineering Energy Storage.
DOI: 10.1016/B978-0-12-814100-7.00009-2
Copyright © 2017 Elsevier Inc. All rights reserved.

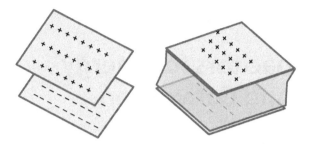

Figure 9.1 A charged conventional capacitor in vacuum (left) and one with a dielectric material (pink (mid gray in print versions)) separating the two electrode plates (right).

The potential is then given as

$$E = \frac{Q \cdot d}{\epsilon_0 \epsilon_i A},\tag{9.3}$$

where ϵ_0, ϵ_i, and A are the dielectric constant of vacuum, the dielectric constant of the body between the parallel plates, and the cross-sectional area. The dielectric constant in vacuum is the lowest dielectric constant. If the volume between the two plates truly were almost vacuum (e.g., in a satellite in space), then ϵ_i would become ϵ_0. For all practical purposes, a spacer is needed to prevent physical contact and short circuit discharge. Thus, the dielectric material must also be electrically insulating. The capacitor plate pair is depicted with (right) and without (left) a dielectric material in Fig. 9.1. The foremost reason for this body to be embedded between the two plates is for spacing or separation since the material inevitably lowers the potential for any given charge Q. However, when taking a charged conventional capacitor with some dielectric material and replacing its dielectric body with a dielectric constant of ϵ_1 by another material with a dielectric constant of ϵ_2, something very interesting happens! The potential changes accordingly:

$$\Delta E = \frac{Q \cdot d}{\epsilon_0 A}\left(\frac{1}{\epsilon_2} - \frac{1}{\epsilon_1}\right).\tag{9.4}$$

Perhaps this change does not appear so interesting at first, but there are two ways to change the potential of an already charged supercapacitor at open circuit, replacing the dielectric material or changing the distance between the plates. If there is air or some other gas between the plates, then the potential changes accordingly:

$$\Delta E = \frac{Q \cdot (d_2 - d_1)}{\epsilon_0 \epsilon_i}. \tag{9.5}$$

Again, for practical reasons, this exercise would start with two plates with a dielectric solid material in between the plates and end with the plates separated by both air and dielectric material. In this instance, the new dielectric would be some combination of a new distance and dielectric constant, where $d_2 > d_1$ and $\epsilon_2 < \epsilon_1$ From Eqs. (9.4) and (9.5) we see that both contributions increase the potential without changing the charge.

Example 9.1: The Capacitive Expansion Effect from a Dielectric Change.

As mentioned in the text, one replaces one dielectric material by another with different dielectric constant while the capacitor is charged at open circuit. In turn, the capacitor will change its potential. This is important later because replacing the electrolyte of a supercapacitor has a very similar effect. I both cases, this can be referred to as an expansion effect; let us study this a little closer.

Consider one polymer dielectric film and a second consisting of the same material but of a porous type such that the dielectric constants relate by $\epsilon_1 = 2\epsilon_2$. When at charge Q, what distance change would this potential change correspond to. How can d_1 and d_2 be related to ϵ_1 and ϵ_2?

Solution:

From Eqs. (9.4) through (9.5) we obtain the potential change:

$$\frac{Q \cdot d_1}{\epsilon_0 A} \left(\frac{1}{\epsilon_2} - \frac{1}{\epsilon_1} \right) = \frac{Q \cdot (d_2 - d_1)}{\epsilon_0 \epsilon_1 A},$$

$$d_1 \left(\frac{2}{\epsilon_1} - \frac{1}{\epsilon_1} \right) = \frac{(d_2 - d_1)}{\epsilon_1},$$

$$d_1 = d_2 - d_1,$$

$$d_1 = \frac{d_2}{2},$$

$$\epsilon_1 = 2\epsilon_2.$$

This means that doubling the dielectric constant is equivalent to lowering the distance by a factor of two.

This is important! Lowering the dielectric constant of the material increases the potentil, similar to increaseing the distance, i.e. lowering the dielectric constant at constant charge makes the capacitor plates experience each other more remote. In other words, when introducing a dielectric material (increasing the dielectric constant real-

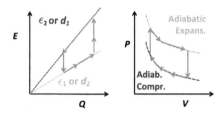

Figure 9.2 Capacitor charging cycles displaying electric work and nonelectric work (left) and the thermodynamic Otto pv-cycle (right), where either heat is exchanged, or adiabatic compression/expansion work of a piston. The illustrations refer to Example 9.1.

tive to vacuum), the plates appear closer, and they have a lower potential relative to the absolute charge readily in the system!

Remark: Changing the distance d or the dielectric material (effectively, the ϵ_i), will require mechanical work equivalent to $-F\Delta E$. Thus the potential can change without a change in the electric charge. One could now cycle a capacitor by charging it and change the dielectric material or thickness at constant charge (using mechanical energy) and then discharge the capacitor and have a larger amount electric work returned. This is illustrated in Fig. 9.2. This is analogous to the Otto pv-cycle; work is added in an adiabatic compression, then heat is added at no change in work, next, there is adiabatic expansion before the heat engine is cooled to the starting value. Perhaps these examples appear misplaced, but once we turn to supercapacitors, they will come in very handy!

9.2 SUPERCAPACITORS

Supercapacitors can store several tenfolds of charge per weight and volume compared to conventional dielectric plate capacitors, and hence the term supercapacitor. To do so, one takes advantage of the capacitive effect that naturally occurs between a solid surface and a liquid with ions or polarizable components. This effect is a sum of two capacitive effects and is therefore referred to as the double layer effect [87], which is described in detail in this section.

Moreover, the material typically used for the double layer charging is a highly porous and nonreactive material. The most commonly used material is activated porous carbon because the porosity is high and the reactivity low; however, transition metals are also sometimes used [88]. The carbon particles are often bound together using some polymeric binder and then submerged in water or an organic electrolyte. Porosity in this context is needed because one wants a large active surface area per weight and volume.

Several efforts are currently made to increase the electronic conductivity and the active area, like adding carbon nanotubes, onion like carbon, or even graphene [89]. The capacitive processes taking place on the surface crumbled together inside the carbon structures are described in the following.

A classic textbook example describing the charging/discharging processes taking place uses the double layer of a flat surface in contact with a saline aqueous solution, for example, NaCl. The anions tend to naturally adsorb on the electrode surface. This layer is referred to as the inner layer. In the outer layer, *the diffuse layer*, the cations accumulate to obtain electroneutrality, though this layer is not as distinct as the inner, and hence its name [90]. This is illustrated in Fig. 9.3A. The inner layer is often referred to as the Stern layer, and the outer diffuse layer as the Gouy–Chapman layer. The inner layer has properties similar to a very thin (Ångstrøm) conventional capacitor at vacuum, whereas the diffuse layer can be of several nanometers.

The theory of the double-layer capacitance and its potential over its layers is described by the Gouy–Chapman–Stern theory, GCS, which is derived from the Poisson–Boltzmann equation. The Poisson–Boltzmann equation is a second-order differential equation expressing the electrochemical potential in a double layer, which from a three-dimensional form $\nabla^2 E$ simplifies to a one-dimensional form $\frac{\partial^2 E}{\partial x^2}$:

$$\frac{\partial^2 E}{\partial x^2} = \frac{Q}{\epsilon_0 \epsilon_i A d},\tag{9.6}$$

where Q/Ad is the local charge per volume, and ϵ_0 and ϵ_i are the dielectric constants. This theory explains the accumulation of charge in the layer Q as a function of temperature T and molar salinity $\bar{c}$;

$$Q = -\sqrt{8\bar{c}N_A\epsilon_0\epsilon_i k_B T} \sinh\left(\frac{e\,E}{2k_B T}\right),\tag{9.7}$$

where N_A, k_B, and e are the Avogadro number, Boltzmann's constant, and the charge of an electron, respectively. From the *Poisson–Boltzmann* equation the potential for the Gouy–Chapman as a function of temperature and salinity can be obtained analytically [90]. Replacing $k_B N_A$ with the molar gas law constant $\bar{R}$ and $N_A e$ with Faraday's constant F, we obtain:

$$E = \frac{2\bar{R}T}{F}\sinh^{-1}\left(\frac{Q}{\sqrt{8FT\,\epsilon_0\epsilon_i\bar{c}}}\right)\tag{9.8}$$

or $E = \dfrac{2\bar{R}T}{F} \ln\left(\dfrac{Q}{\sqrt{\xi\bar{c}}} + \sqrt{\dfrac{Q^2}{\xi\bar{c}} + 1}\right)$, where $\xi = \sqrt{8FT\,\epsilon_0\epsilon_r}$.   (9.9)

In turn, under the constraint $Q^2\xi^{-1} \gg \bar{c}$, this can be rewritten as

$$E = -\dfrac{2\bar{R}T}{F}\left(\ln\bar{c} + \ln\left[\xi^{1/2}Q\right]\right),$$ (9.10)

$$\Delta E_{1-2} = \dfrac{2\bar{R}T}{F}\left(\ln\dfrac{\bar{c}_1}{\bar{c}_2}\right),$$ (9.11)

which is an expression that not only fits very well with experimental observations, but also fits with the concentration relation of the Nernst equation and the entropic mixing term of Eq. (2.40) (p. 26), when saying that concentration change and activity change are the same.

Equation (9.11) shows how the potential of the capacitive system changes with the salt concentration $\bar{c}$, that is, the potential increases when dilute solution enters the system and decreases when concentrated saline solution enters the system. As shown in Example 9.2, this equation can be used for understanding concentration and it relation to the potential when charge is kept constant.

Example 9.2: Concentration and Potential Effects at Open Circuit.
Considering that doubling the concentration ($\bar{c}_2 = 2\bar{c}_1$) does not significantly change the dielectric constant of the solution $\epsilon_{sol.}$. How can the concentration be linked to any change in a conventional thin-film capacitor in vacuum? (Compare Eq. (9.11) to Eq. (9.5)).
Answer:
The potential changes ΔE_{1-2} are described as follows only due to the second term in Eq. (9.11):

$$\dfrac{Q\cdot(d_2 - d_1)}{\epsilon_0\epsilon_{sol.}} = \dfrac{\bar{R}T}{F}\ln\left[\dfrac{\bar{c}_1}{\bar{c}_2}\right],$$

$$d_2 - d_1 = \dfrac{\epsilon_0\epsilon_{sol.}\bar{R}T}{QF}\ln 0.5 < 0.$$

When doubling the concentration, the potential change is negative ($\ln 0.5$). This corresponds to the effect of decreasing distance in a conventional capacitor. The other way around is that, when at constant charge, moving from a higher concentration to a lower concentration, the capacitor interdistance appears increased. Because a super-

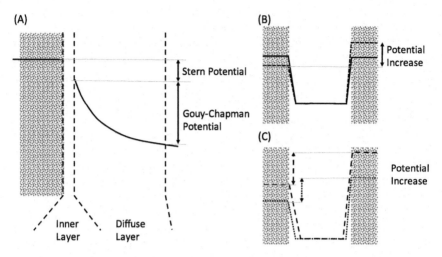

Figure 9.3 A possible potential distribution across a double layer of a flat electrode and into the adjacent electrolyte region (A), e.g., in water. The potential increases in the double layers when externally charged (positively on one electrode and negatively on the opposite) (B), and further when lowering the bulk concentration (C) (after first having charged the electrode pair).

capacitor has double layers, the effect of the potential increase when going to a lower concentration is called the double layer expansion effect!

Possible double layer scenarios are shown in Fig. 9.3. In (A), the double layer and a possible potential distribution of the diffuse layer are shown for one electrode–electrolyte interface. The potential distribution of the double layer of the electrode changes with the solution concentration $\bar{c}$ and the charge applied to the double layer Q (Eq. (9.9)). In (B), the two-electrode capacitor is charged externally, and one double layer lowers its potential while that of the other electrode increases. When the two electrodes are charged, so that they would have different potentials, the double layers both expand if the concentration in the bulk is lowered (C). Thus the total potential increases.

Following the analogy of Examples 9.1 and 9.2, we can propose a cycle that harvests the energy of mixing sea and river water, that is, salinity gradient energy [4]. By first charging two electrodes in sea water, next, change the water to river water at open circuit so that the double layer expansion will ramp up the potential. (The energy for this effect is supplied by the chemical potential difference between salt in sea and river water.) We can third discharge the capacitors and retrieve more work than we implemented in the first step. Eventually, we need to replace the river water with sea water at discharged state. This technology is called capacitive power extraction (CPE). This technology can also be used for desalination when operated in the reverse way; then it is known as capacitive desalination or capacitive deionization (CDI). Combining the two can be used for energy storage, using first CDI to build two gigantic reservoirs of different salinity and later using CPE to get the electric energy back, similarly as for ED and RED (see Kingsbury et al. [19]).

So far CPE, CDI, and the combination battery as a technology are only very briefly explored and reported in the literature [91,92]. Still, the technology seems promising in terms of impeding several kinds of fouling mechanisms reported for RO, FO, PRO, ED, and RED (several different desalination or salt power technologies) systems. This is due to the stress imposed on many organisms when exposed to both sea and river water and also because scaling from river water may be cleansed by sea water and vice versa. Further, the charging and discharging of the double layer is a reversible process, so that energy losses such as in a turbine or for an electrochemical reaction are absent. A disadvantage of the CPE techniques is that the electricity they produce is not only dynamic, but also alternating. Hence auxiliary systems to aid the CPE to deliver a steady DC current are required.

For sodium chloride solutions, if one coulomb is removed from an electrode, and in the absence of electrochemical reactions, a mole of anions Cl^- will adsorb the surface (the inner layer). Correspondingly, a surplus of a mole cations Na^+ will be present in the diffuse layer [93]. Thus the electroneutrality is fulfilled between these two layers and for the electrode. The process can of course be reversed. Introducing NaCl to a diluted aqueous solution will thus by this manner make it energetically favorable for the electrode to get rid of electrons. Due to the relative large specific surface area of some electrode materials, typically active carbon or carbon black, the electrodes have the ability to adsorb and desorb relatively large amounts of ions and thus store electrons and vacancies.

Moreover, the examples here are explained using sodium chloride in water because this is a solution that most readers are familiar with. However, because water starts to split into hydrogen and oxygen above cell potentials above 1.23 V, one often applies organic solutions like e.g. 1 M solution of tetraethylammoniume tetrafluoroborate (TEA-BF$_4$), in acetonitrile (ACN) [94] in commercial supercapacitors.

Table 9.1 Calculation rules for resistors and capacitors in series and parallel

	Series	Parallel
Capacitors	$C_{tot}^{-1} = \sum_{i=1}^{n} \frac{1}{C_i}$	$C_{tot} = \sum_{i=1}^{n} C_i$
Resistance	$R_{tot} = \sum_{i=1}^{n} R_i$	$R_{tot}^{-1} = \sum_{i=1}^{n} \frac{1}{R_i}$

9.3 DEPLOYING SUPERCAPACITORS

When deploying supercapacitors for energy storage, the most important drawback to account for is the self-discharge that occurs after a few days, if not even only a few hours. The mechanism is at its simplest, explained as when a supercapacitor is charged and left at rest, the charge and consequently also the potential will gradually be discharged internally. This happens transiently, like the water volume and height in the classical engineering example of the barrel with a hole in the bottom. This goes well along with the explanation that the charge accumulation in the diffuse layer diffuses into the electrolyte, driven more strongly at higher potential. The discharge occurs mainly via two mechanisms, the diffusive force of the diffuse layer and subsequently a charge leakage across the inner layer [95].

As said before, the supercapacitors have enormous capacity in terms of having small systems taking up and delivering electric power. Setting up a set of supercapacitors based on of-the-shelf supercapacitors is in principal a trivial exercise; however, care must be taken when assembling a combination of series and parallels. Based on the need for voltage and energy stored, one can fix supercapacitors in a combined system of parallel and series network. In doing so, one must remember how resistance and capacitors differ in circuit rules. These are listed in Table 9.1.

When charging a supercapacitor, a constant current can be applied, like in Fig. 9.2 (left), and the potential increases linearly. However this is only valid for a capacitor with no internal resistance. Supercapacitors are electrochemical devices and have internal resistance, which is in series with the capacitor just like any other electrochemical cell. This is illustrated in Fig. 9.4. When charging, there is an extra potential for the ohmic loss inside the capacitor. This is illustrated by the cell potential line (dash–dot) being larger than the reversible potential line (solid), which is the potential stored and available for use afterwards. When putting together a combina-

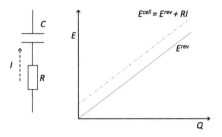

Figure 9.4 The relation between the cell potential and reversible of a supercapacitor in series with its internal resistance.

tion of supercapacitors in series and parallel, the ohmic potential drop must be accounted for since this lowers the potential to the supercapacitor. Getting the right amount of supercapacitors can thus be a little cumbersome, as shown in Example 9.3.

Example 9.3: Supercapacitors in Parallel and Series.

Each wind turbine in a wind power park needs supercapacitors for system control and power buffer of wind fluctuations. The requirements for the supercapacitors are that they must handle a voltage change of 220 to 300 V. The current will be 13 A. The charging and discharging cycles can last for up to 8 seconds. Each of the of-the-shelf supercapacitors tolerate up to 2.7 V and has a maximum capacity C_i of 50 F and an internal resistance of 5 mΩ.

a) What is the capacitance needed for the system?

b) The supercapacitors are coupled in series to tolerate 300 V. What is the capacitance for this series of supercapacitors?

c) How many series like the one above must be in parallel to provide the needed capacitance?

d) Considering that the maximum current is applied at the highest applied voltage, what is then the voltage that remains stored in such a circuit? (3 parallels of 112 supercapacitors in series.)

e) What is the energy efficiency at this point?

Solution:

a) *From Eq. (9.1) we can calculate the system total capacitance:*

$$C = \frac{\Delta Q}{\Delta E} = \frac{I \Delta t}{\Delta E} = 13 \text{ [A]} \frac{8 \text{ [s]}}{80 \text{ [V]}} = 1.30 \text{ F.}$$

b) *When maximally charged and in series, we need $n = \frac{300 \text{ [V]}}{2.7 \text{ [V]}} = 111.1$ supercapacitors, or in fact 112 supercapacitors to not exceed the 2.7 V limit. The capaci-*

tance of such a series is (from Table 9.1)

$$C_{tot} = \left(\frac{112}{50\ [F]}\right)^{-1} = 0.446\ [F].$$

c) *Capacitance in parallel is linear, so when the capacitance needed is 1.3 F, three parallels ($C = 1.34$ F) will be sufficient. So each of the serially coupled capacitors has capacitance 0.446 F, two lines will have 0.892 F, and three in parallel have 1.338 F.*

d) *First, we must calculate the resistance of the circuit:*

$$R_{tot} = \left(\frac{3\ parallels}{R_{series}}\right)^{-1} = \left(\frac{3\ parallels}{112 \cdot R_{single}}\right)^{-1}$$

$$= \left(\frac{3}{112 \cdot 0.005[\Omega]}\right)^{-1} = 0.19\ [\Omega].$$

Next, we calculate the reversible potential when the supercapacitors meet the limiting voltage: $E^{rev} = E^{cell} - IR = 300\ [V] - 0.19\ [\Omega] \cdot 13\ [A] = 297.6\ [V].$

e) *The energy efficiency becomes $\varepsilon = \frac{E^{cell}}{E^{rev}} = \frac{297.4\ [V]}{300\ [V]} = 99.2\%.$*
One remark is that the resistance of each capacitor is very low. However, even if assuming a $0.01\ \Omega$ resistance in the connectors between any supercapacitors, in turn, leading to an increase in the ohmic resistance of a factor of three, would give an efficiency of 97.6%. Compared to a battery, it is partly the absence of a Tafel overpotential that leads to such very high energy efficiencies.

9.4 PSEUDO- AND HYBRID SUPERCAPACITORS

One disadvantage of conventional capacitors is the low energy capacity. Thus the supercapacitors where developed. When evaluating the energy efficiency and energy storage capacity of these, they have a high energy efficiency when nearly fully charged and still rather low energy capacity. To make them store more energy, several researchers have started to add red-ox active components to the supercapacitors. This is often the lighter transition metal with partly filled d-orbitals in oxide forms (e.g., Mn_2O_3 [96] or Fe_2O_3 [97]). Such supercapacitors are hybrids between a red-ox battery and a supercapacitor and are often referred to as hybrid capacitors or pseudo-capacitors. Applying such capacitors is in many ways similar to applying a battery. This is because only one of the electrodes has this red-ox layer, whereas a regular supercapacitor has two equal electrodes. Since pseudo- or hybrid capacitors are more battery like, they are not further discussed here.

PROBLEMS

Problem 9.1. Comparing Technologies.

The current density of a supercapacitor under heavy duty is of order $200\ \mathrm{A\,m^{-2}}$. The conductivity is very similar to a battery in Fig. 7.7. The thermal conductivity is of order $0.4\ \mathrm{W\,K^{-1}\,m^{-1}}$ for the electrodes and $0.2\ \mathrm{W\,K^{-1}\,m^{-1}}$ [94] for the separator, and the somewhat higher for a Li–ion battery (LiB) [39,98]. An MEA of a PEMFC also has a thermal conductivity of around $0.2\ \mathrm{W\,K^{-1}\,m^{-1}}$ [71,99].

a) Create an overview of relevant current densities, ohmic potential losses, component thicknesses, and electrode thicknesses for Li–ion batteries, supercapacitors, and a PEMFC.

b) How are these devices assembled up to a point where cooling can be applied?

c) How does this affect the internal temperature gradients in the devices?

d) Develop relevant expressions for the different technologies and determine relevant temperature differences.

SOLUTIONS

Solution to Problem 9.1. Comparing Technologies.

a) We now sum up an estimate of the properties electrochemical energy storage devices have:

Technology	j $\mathrm{A\,m^{-2}}$	$\delta_{electrode}$ µm	$k_{electrode}$ $\mathrm{W\,K^{-1}\,m^{-1}}$	$\delta_{electrolyte}$ µm	$k_{electrolyte}$ $\mathrm{W\,K^{-1}\,m^{-1}}$	κ $\mathrm{S\,m^{-1}}$	η V
LiB	20–100	70–120	0.5–0.9	20–30	0.2–0.3	0.10	0.05–0.10
S.C.	100–400	50–100	0.3–0.5	20–30	0.2–0.3	0.10	–
PEMFC	5000–20 000	200–400	0.2–0.4	30–50	0.2–0.3	10	0.4–0.5

b) Supercapacitors are assembled very similarly to Li–ion batteries, either stacked or rolled up. This means that there are several layers of heat generating zones inside them and cooling can be applied on the surface. A PEMFC typically has cooling either in the gas channels or inside the gas feed plates.

c) These designs means that supercapacitors and Li–ion batteries have several layers of heat generating and nongenerating layers that alternate. If we consider these flat plates, one approach is to consider several millimeters of a uniformly distributed heat generation. This gives a parabolically shaped temperature gradient inside these devices. When evaluating a rolled up configuration, however, the parabolic shape flattens and lowers the gradients for two reasons, the geometry factor and the fact that heat is conducted via the current collectors in a spiral shape to the outer surface.

For the PEM fuel cell, it has its heat source in the middle of the so-called GDLs (gas diffusion layers), and the temperature profile in these layers is linear throughout the GDL as a first-order approximation.

d) For the layered supercapacitors and Li-ion batteries, we can take Fourier's second law of heat transport in one direction:

$$\frac{\partial T^2}{\partial^2 x} k_{eff} = \frac{\dot{Q}}{V},$$

where

$$k_{eff} = \frac{\delta_{tot}}{\sum \frac{\delta_i}{k_i}}$$

is the effective thermal conductivity, and

$$\frac{\dot{Q}}{V} = \frac{\dot{q}}{\delta_{tot}} = \frac{\frac{\delta_i}{\kappa_i} j^2 + \eta j}{\delta_{tot}}.$$

Using the boundary conditions that the temperature gradient is zero in the middle (and consider the middle being $x = 0$) and that we have a fixed temperature at the surfaces, we can obtain the maximum temperature difference for the two devices:

$$\Delta T_{max} = \frac{\dot{q}}{\delta_{tot} k_{eff}} \frac{(n\delta_{tot})^2}{8},$$

where n reflects the amount of layers of single cells.

For the fuel cell, revisiting Example 8.3 tells us that the equivalent equation would be

$$\Delta T_{max} = \frac{\dot{q}}{2} \frac{\delta_{GDL}}{k_{GDL}}.$$

Thus we can instate this information into the following table:

Technology	$\dot{q}$ /W m^{-2}	δ_{tot} μm	k_{eff} /W K^{-1} m^{-1}	n	ΔT_{max} /K
LiB	13	6170	0.40	29	0.6
S.C.	48	130	0.30	38	3.8
PEMFC	14000	650	0.29	1	3.9

In this table, the thinnest electrodes and thickest membranes are considered along with the most intense current densities. For the LiB and the S.C., a total thickness of 5 mm was considered, and the current collectors neglected. Nevertheless, from the estimates in this problem, it can be concluded that both heat management and internal temperature gradients should be considered in electrochemical energy storage devices. This is an engineering task that is likely to become a lot more relevant in the future since renewable energy needs to be stored and cars, buses, and ships need to be charged in a short time.

APPENDIX A

Symbols and Constants

ROMAN LETTERS

Symbol	Units	Description
a	V	Tafel constant
b	V/ord. of magn.	Tafel slope
C	J/K	Heat capacity
c	J/K kg	Specific heat capacity
c_i	kg/m^3	Concentration of gas or component i
$\bar{c}$	J/K mole	Molar heat capacity
$\bar{c}_i$	mole/L	Molar concentration of component i
$\hat{c}$	$J/K\,m^3$	Volumetric heat capacity
F	C/mole	Faraday constant (96 485)
G	J	Gibbs free energy
g	m/s^{-2}	Gravitational acceleration constant.
$\bar{g}$	J/mole	Gibbs free molar energy
H	J	Enthalpy
h	J/kg	Specific enthalpy
$\bar{h}$	J/mole	Molar enthalpy
h	m	Head in hydraulic systems
m	kg	Mass
$\dot{m}$	kg/s	Mass rate
P	W or J/s	Power
p	Pa, atm, or bar	Pressure
p_i	Pa, atm, or bar	Partial pressure of component i
R	J/K kg	Specific gas constant
$\bar{R}$	J/K mole	Molar gas constant (8.314)
S	J/K	Entropy
$\bar{s}$	J/K mole	Molar entropy
T	K or °C	Temperature
t	s or h	Time
U	J	Internal energy
Q	J	Heat (thermal energy to the system)
Q	C or coloumb	Charge (electric charge)
$\dot{Q}$	W	Heat to the system (per time)
V	m^3	Volume
v	m^3/kg	Specific volume
v	m/s	Velocity
W	J	Work from the system
$\dot{W}$	W	Power from a control volume
w	J/kg	Specific work
$\dot{w}$	W/kg	Specific power
z	equiv./mole	Moles of electrons per mole reactant

GREEK LETTERS

Symbol	Units	Description
α	–	Symmetry coefficient in the Butler–Volmer equation
α_i	–	Apparent membrane permselectivity
ϵ_i	F/m	Dielectric constant of medium i
ε	– or %	Efficiency
γ	dm^3/mole	Molar activity coefficient
δ	μm	Distance between electrodes
κ	S/m	Conductivity
Λ	$S\,cm^3$/mole m	Molar specific conductivity
μ_i	J/mole	Chemical potential of component i
Ω	ohm	Electric resistance
ρ	kg/m^3	Density

CONSTANTS

Symbol	Value	Units	Description
F	96485	C equiv.$^{-1}$	Faraday's Constant
g	9.81	m/s^2	Gravimetric acceleration on earth
$\bar{R}$	8.314	J/mole K	Ideal gas law constant

APPENDIX B

Adiabatic Compression of Air

$T/$ K	$T/$ °C	v_r	$u/$ kJ kg^{-1}	p_r	$h/$ kJ kg^{-1}	$s^o/$ J kg^{-1} K^{-1}
200	−73	270	143	0.248	200	1.30
210	−63	239	150	0.294	210	1.34
220	−53	213	157	0.346	220	1.39
230	−43	191	164	0.404	230	1.44
240	−33	172	171	0.469	240	1.48
250	−23	155	178	0.541	250	1.52
260	−13	140	185	0.621	250	1.56
270	−3	128	193	0.708	270	1.60
280	7	117	200	0.804	280	1.63
285	12	112	203	0.855	285	1.65
290	17	107	207	0.909	290	1.67
295	22	103	210	0.909	290	1.69
298	25	100	213	1	298	1.70
300	27	98.3	214	1.02	300	1.70
305	32	94.3	218	1.08	305	1.72
310	37	90.6	221	1.15	310	1.73
315	42	87.0	225	1.21	315	1.75
320	47	83.7	228	1.28	320	1.77
325	52	80.5	232	1.35	325	1.78
330	57	77.4	236	1.43	330	1.80
340	67	71.9	243	1.59	340	1.83
350	77	66.8	250	1.76	350	1.86
360	87	62.3	257	1.94	361	1.89
370	97	58.1	264	2.14	371	1.91
380	107	54.3	272	2.35	381	1.94
390	117	50.9	279	2.57	391	1.97
400	127	47.7	286	2.57	391	1.99
410	137	44.8	293	3.07	411	2.02
420	147	42.2	301	3.34	421	2.04
430	157	39.7	308	3.63	431	2.07
440	167	37.5	315	3.94	442	2.09
450	177	35.4	323	4.26	452	2.11

continued on next page

$T/$ K	$T/$ °C	v_r	$u/$ kJ kg^{-1}	p_r	$h/$ kJ kg^{-1}	$s^o/$ J kg^{-1} K^{-1}
460	187	33.5	330	4.61	462	2.13
470	197	31.7	337	4.98	472	2.16
480	207	30.0	345	5.37	482	2.18
490	217	28.4	352	5.78	493	2.20
500	227	27.0	359	6.21	503	2.22
510	237	25.7	367	6.67	513	2.24
520	247	24.4	374	7.15	524	2.26
530	257	23.2	382	7.66	534	2.28
540	267	22.1	389	8.20	544	2.30
550	277	21.1	397	8.76	556	2.32
560	287	20.1	404	9.35	565	2.34
570	297	19.2	412	9.97	576	2.36
580	307	18.3	420	10.6	586	2.37
590	317	17.5	427	11.3	597	2.39
600	327	16.7	435	11.3	597	2.41
610	337	16.0	442	12.8	618	2.43
620	347	15.3	450	13.6	628	2.44
630	357	14.7	458	14.6	639	2.46
640	367	14.1	466	15.2	649	2.48
650	377	13.5	473	16.1	660	2.49
660	387	13.0	481	17.1	670	2.51
670	397	12.4	489	18.1	681	2.53
680	407	11.9	497	19.1	692	2.54
690	417	11.5	504	20.2	703	2.56
700	427	11.0	512	21.3	713	2.57
710	437	10.6	520	22.4	724	2.59
720	447	10.2	528	23.6	735	2.60
730	457	9.83	536	24.9	746	2.62
740	467	9.47	544	26.2	756	2.63
750	477	9.12	552	27.6	767	2.65
760	487	8.79	560	29.0	778	2.66
780	507	8.17	576	32.0	800	2.69
800	527	7.61	592	35.3	822	2.72
820	547	7.10	609	38.8	844	2.75
840	567	6.62	625	42.5	866	2.77
860	587	6.19	641	46.6	888	2.80
880	607	5.79	658	50.9	911	2.82
900	627	5.43	675	55.6	933	2.85
920	647	5.09	691	60.6	955	2.87

continued on next page

$T/$ K	$T/$ °C	v_r	$u/$ kJ kg^{-1}	p_r	$h/$ kJ kg^{-1}	$s^o/$ J kg^{-1} K^{-1}
940	667	4.78	708	65.9	978	2.90
960	687	4.49	725	71.6	1001	2.92
980	707	4.23	742	77.7	1023	2.94
1000	727	3.98	759	84.2	1046	2.97
1020	747	3.75	776	91.1	1069	2.99
1040	767	3.69	793	98.4	1092	3.01
1060	787	3.35	811	106	1115	3.03
1080	807	3.16	828	115	1138	3.06
1100	827	2.99	845	123	1161	3.08
1120	847	2.83	863	133	180	3.10
1140	867	2.68	880	143	1208	3.12
1160	887	2.54	898	153	1231	3.14
1180	907	2.41	916	164	1254	3.16
1200	927	2.29	933	176	1278	3.18
1220	947	2.18	951	188	1301	3.20
1240	967	2.07	969	201	1325	3.22
1260	987	1.97	987	215	1349	3.24
1280	1007	1.87	1005	229	1372	3.26
1300	1027	1.78	1023	244	1396	3.27
1320	1047	1.70	1041	260	1420	3.29
1340	1067	1.55	1077	277	1444	3.33
1360	1087	1.62	1059	295	1467	3.31
1380	1107	1.48	1095	313	1491	3.34
1400	1127	1.41	1114	333	1515	3.36
1420	1147	1.35	1132	353	1539	3.38
1440	1167	1.23	1168	374	1564	3.41
1460	1187	1.29	1150	397	1588	3.40
1480	1207	1.18	1187	420	1612	3.43
1500	1227	1.13	1205	444	1636	3.45
1520	1247	1.08	1224	470	1660	3.46
1540	1267	1.04	1242	497	1685	3.48
1560	1287	1.00	1261	525	1709	3.49
1580	1307	0.96	1280	554	1733	3.51
1600	1327	0.918	1298	584	1758	3.52
1620	1347	0.882	1317	616	1782	3.54
1660	1387	0.815	1354	683	1831	3.57
1680	1407	0.783	1373	719	1856	3.58
1700	1427	0.753	1393	757	1880	3.60
1750	1477	0.685	1440	857	1942	3.63

continued on next page

$T/$ K	$T/$ °C	v_r	$u/$ kJ kg^{-1}	p_r	$h/$ kJ kg^{-1}	$s^o/$ J kg^{-1} K^{-1}
1850	1577	0.570	1535	1089	2065	3.70
1900	1627	0.521	1583	1222	2127	3.74
1950	1677	0.478	1631	1367	2190	3.77
2000	1727	0.439	1679	1527	2252	3.80
2100	1827	0.373	1775	1890	2378	3.86
2150	1877	0.344	1824	2095	2440	3.89
2200	1927	0.318	1872	2317	2503	3.92
2250	1977	0.295	1921	2558	2566	3.95
2100	1827	0.373	1775	1890	2378	3.86
2150	1877	0.344	1824	2095	2440	3.89
2200	1927	0.318	1872	2317	2503	3.92
2250	1977	0.295	1921	2558	2566	3.95

Originally published in J. H. Keenan and J. Kaye, Gas Tables (New York: John Wiley & Sons, 1948)

APPENDIX C

Para- and Ortho-Hydrogen

Thermodynamic properties of hydrogen at different temperatures, equilibrium content of para-hydrogen, and specific conversion energy from normal hydrogen to para-hydrogen. Republished by McCarty et al. [57].

$T/$ K	Para /%	Δh kJ kg^{-1}	$T/$ K	Para /%	Δh kJ kg^{-1}
10	99.9999	527	90	42.9	500.8
20	99.82	527	100	38.6	482
30	97	527	120	33	427
33	95	527	150	28.6	322
40	88.7	527	200	26	263
50	77	526.8	250	25.3	70.5
60	65.6	525.5	298	25.1	28.6
70	56	521.8	300	25.1	27.6
80	48.5	513.9	500	25	–

BIBLIOGRAPHY

[1] Clean energy's dirty secret. The Economist 25 February 2017.

[2] Aylward GH, Findlay T. SI chemical data. 4 edition. John Wiley & Sons Australia, Ltd.; 1998.

[3] Burheim Odne S, Pharoah Jon G, Vermaas David, Sales Bruno B, Nijmeijer Kitty, Hamelers Hubertus VM. Reverse electrodialysis. Encyclopedia of membrane science and technology; 2013.

[4] Bijmans MFM, Burheim OS, Bryjak M, Delgado A, Hack P, Mantegazza F, Tenisson S, Hamelers HVM. Capmix – deploying capacitors for salt gradient power extraction. Energy Procedia 2012;20:108–15.

[5] Huggins Robert Alan. Energy storage. Springer; 2010.

[6] Courty JM, Kierlik E. Les chaufferettes chimiques. Pour la Science 2008:108–10.

[7] Dincer Ibrahim, Rosen Marc. Thermal energy storage: systems and applications. John Wiley & Sons; 2002.

[8] Førland KS, Førland S, Kjelstrup T. Irreversible thermodynamics. Trondheim: Tapir Academic Press; 2001. Reprint from Jon Wiley and Sons, Inc.

[9] Bradshaw Robert W, Cordaro Joseph G, Siegel Nathan P. Molten nitrate salt development for thermal energy storage in parabolic trough solar power systems. In: ASME 2009 3rd international conference on energy sustainability collocated with the heat transfer and InterPACK09 conferences. American Society of Mechanical Engineers; 2009. p. 615–24.

[10] Key world energy statistics. IEA – public reports, International Energy Agency; 2016.

[11] Weast RC. CRC handbook of chemistry and physics. 58th edition. CRC Press, Inc.; 1977.

[12] Chagnes A, Carr B, Willmann P, Lemordant D. Modeling viscosity and conductivity of lithium salts in γ-butyrolactone. Journal of Power Sources 2002;109:203–13.

[13] Ding Michael S. Electrolytic conductivity and glass transition temperature as functions of salt content, solvent composition, or temperature for lipf$_6$ in propylene carbonate + diethyl carbonate. Journal of Chemical & Engineering Data 2003;48:519–28.

[14] Pharoah JG, Burheim OS. On the temperature distribution in polymer electrolyte fuel cells. Journal of Power Sources 2010;195:5235–45.

[15] Prausnitz JM, Lichtenthaler RN, de Azevedo EG. Molecular thermodynamics of FluidPhase equilibria. 3 edition. Upper Sadle River, New Jersey: Prentice Hall PTR; 1999.

[16] Atkins Peter, de Paula Julio. Physical chemistry. 7 edition. Oxford University Press; 2002.

[17] Galama AH, Daubaras G, Burheim OS, Rijnaarts HHM, Post JW. Fractioning electrodialysis: a current induced ion exchange process. Electrochimica Acta 2014;136:257–65.

[18] Zlotorowicz Agnieszka, Viktor Strand Robin, Stokke Burheim Odne, Wilhelmsen Øivind, Kjelstrup Signe. The permselectivity and water transference number of ion exchange membranes in reverse electrodialysis. Journal of Membrane Science 2017;523:402–8.

[19] Kingsbury Ryan S, Chu Kevin, Coronell Orlando. Energy storage by reversible electrodialysis: the concentration battery. Journal of Membrane Science 2015;495:502–16.

[20] Burheim Odne S, Seland Frode, Pharoah Jon G, Kjelstrup Signe. Improved electrode systems for reverse electro-dialysis and electro-dialysis. Desalination 2012;285:147–52.

[21] Weinstein JN, Leitz FB. Electric power from differences in salinity: the dialytic battery. Science 1976;191:558.

[22] Newman J, Thomas-Alyea KE. Electrochemical systems. 3rd edition. John Wiley et Sons Inc.; 2004.

[23] Długołęcki P, Nijmeijer K, Wessling M. Current status of ion exchange membranes for power generation from salinity gradients. Journal of Membrane Science 2008;319:214–22.

[24] Walsch Frank C. A first course in electrochemical engineering. 1 edition. The Electrochemical Consultancy; 1993.

[25] Vetter J, Novák P, Wagner MR, Veit C, Möller K-C, Besenhard JO, Winter M, Wohlfahrt-Mehrens M, Vogler C, Hammouche A. Ageing mechanisms in lithium-ion batteries. Journal of Power Sources 2005;147(1):269–81.

[26] Richter F, Kjelstrup S, Vie PJS, Burheim O. Measurements of ageing and thermal conductivity in secondary Li-ion batteries and their impact on internal temperature profiles. Electrochimica Acta 2017;XXX.

[27] Waldmann Thomas, Wilka Marcel, Kasper Michael, Fleischhammer Meike, Wohlfahrt-Mehrens Margret. Temperature dependent ageing mechanisms in lithium-ion batteries – a post-mortem study. Journal of Power Sources 2014;262:129–35.

[28] Dustmann Cord-H. Advances in zebra batteries. Journal of Power Sources 2004;127(1):85–92.

[29] Gerssen-Gondelach Sarah J, Faaij André PC. Performance of batteries for electric vehicles on short and longer term. Journal of Power Sources 2012;212:111–29.

[30] Goodenough John B, Park Kyu-Sung. The Li-ion rechargeable battery: a perspective. Journal of the American Chemical Society 2013;135(4):1167–76.

[31] Thomas CE Sandy. Fuel cell and battery electric vehicles compared. International Journal of Hydrogen Energy 2009;34:9279–96.

[32] Gualous Hamid, Gallay Roland. Supercapacitor module sizing and heat management under electric, thermal, and aging constraints. chapter 11. Wiley–VCH Verlag GmbH & Co. KGaA; 2013. p. 373–476.

[33] Bujlo P, Xie CJ, Shen D, Ulleberg O, Pasupathi S, Pasciak G, Pollet BG. Hybrid polymer electrolyte membrane fuel cell–lithium-ion battery powertrain testing platform – hybrid fuel cell electric vehicle emulator. International Journal of Energy Research 2017. http://dx.doi.org/10.1002/er.3736.

[34] Ressem AO, Taksdal K, Stakvik R. The green transition of the bus fleet in Oslo and Akershus. Bachelor thesis. Norway: NTNU; 2016.

[35] Wagner Frederick T, Lakshmanan Balasubramanian, Mathias Mark F. Electrochemistry and the future of the automobile. Journal of Physical Chemistry Letters 2010;1(14):2204–19.

[36] Department of Oil, USA Energy. Fuel cell technologies office information resources. http://energy.gov/eere/fuelcells/fuel-cell-technologies-office-information-resources.

[37] Renault, EVObsession. Electric car pollution much less than gas or diesel car pollution. http://evobsession.com/electric-car-pollution-much-less-than-gas-or-diesel-car-pollution/.

[38] Sakaebe Hikari. Zebra batteries. In: Encyclopedia of applied electrochemistry. Springer; 2014. p. 2165–9.

[39] Stokke Burheim Odne, Onsrud Morten Andreas, Pharoah Jon G, Vullum-Bruer Fride, Vie Preben JS. Thermal conductivity, heat sources and temperature profiles of Li-ion batteries. ECS Transactions 2014;58(48):145–71.

[40] Zhang Zheng, Wang Hui, Ji Shan, Pollet Bruno G, Wang Rongfang. V2o5-sio2 hybrid as anode material for aqueous rechargeable lithium batteries. Ionics 2016;22(9):1593–601.

[41] Tobishima S. Thermal runaway. In: Secondary batteries – rechargeable systems – lithium-ion; 2009. p. 409–17.

[42] Satyavani TVSL, Kumar A Srinivas, Rao PSV Subba. Methods of synthesis and performance improvement of lithium iron phosphate for high rate Li-ion batteries: a review. Engineering Science and Technology, an International Journal 2016;19(1):178–88.

[43] Wood David L, Li Jianlin, Daniel Claus. Prospects for reducing the processing cost of lithium ion batteries. Journal of Power Sources 2015;275:234–42.

[44] Weber Adam Z, Mench Matthew M, Meyers Jeremy P, Ross Philip N, Gostick Jeffrey T, Liu Qinghua. Redox flow batteries: a review. Journal of Applied Electrochemistry 2011;41(10):1137.

[45] Zhang Jianlu, Li Liyu, Nie Zimin, Chen Baowei, Vijayakumar M, Kim Soowhan, Wang Wei, Schwenzer Birgit, Liu Jun, Yang Zhenguo. Effects of additives on the stability of electrolytes for all-vanadium redox flow batteries. Journal of Applied Electrochemistry 2011;41(10):1215–21.

[46] Aaron DS, Liu Q, Tang Z, Grim GM, Papandrew AB, Turhan A, Zawodzinski TA, Mench MM. Dramatic performance gains in vanadium redox flow batteries through modified cell architecture. Journal of Power Sources 2012;206:450–3.

[47] Divisek J. Water electrolysis in a low-and medium-temperature regime. chapter 2In: Electrochemical hydrogen technologies-electrochemical production and combustion of hydrogen. Oxford: Elsevier; 1990. p. 137–212.

[48] Springer Thomas E, Zawodzinski TA, Gottesfeld Shimshon. Polymer electrolyte fuel cell model. Journal of the Electrochemical Society 1991;138(8):2334–42.

[49] Rasten Egil. Electrocatalysis in water electrolysis with solid polymer electrolyte. PhD-dissertation. NTNU; 2001.

[50] Thomas Owen D, Soo Kristen JWY, Peckham Timothy J, Kulkarni Mahesh P, Holdcroft Steven. A stable hydroxide-conducting polymer. Journal of the American Chemical Society 2012;134(26):10753–6.

[51] Weber André, Ivers-Tiffée Ellen. Materials and concepts for solid oxide fuel cells (sofcs) in stationary and mobile applications. Journal of Power Sources 2004;127(1):273–83.

[52] Dalgaard Ebbesen Sune, Mogensen Mogens. Electrolysis of carbon dioxide in solid oxide electrolysis cells. Journal of Power Sources 2009;193(1):349–58.

[53] Laguna-Bercero MA, Skinner SJ, Kilner JA. Performance of solid oxide electrolysis cells based on scandia stabilised zirconia. Journal of Power Sources 2009;192(1):126–31.

[54] Truls Norby. Solid-state protonic conductors: principles, properties, progress and prospects. Solid State Ionics 1999;125:1–11.

[55] Norby Truls, Widerøe Marius, Glöckner Ronny, Larring Yngve. Hydrogen in oxides. Dalton Transactions 2004:3012–8.

[56] Janz GJ, Neuenschwander E, Kelly FJ. High-temperature heat content and related properties for li2co3, na2co3, k2co3, and the ternary eutectic mixture. Transactions of the Faraday Society 1963;59:841–5.

[57] McCarty Robert D, Hord J, Roder HM. Selected properties of hydrogen (engineering design data). Technical report, Boulder, CO (USA): National Engineering Lab. (NBS); 1981.

[58] Beckmann Michael. Linde hydrogen fueling overview. https://www.hydrogen. energy.gov/pdfs/htac_nov14_9_beckman.pdf, 16 November 2016. p. 12.

[59] Lototskyy Mykhaylo V, Tolj Ivan, Wafeeq Davids Moegamat, Klochko Yevgeniy V, Parsons Adrian, Swanepoel Dana, Ehlers Righardt, Louw Gerhard, van der Westhuizen Burt, Smith Fahmida, et al. Metal hydride hydrogen storage and supply systems for electric forklift with low-temperature proton exchange membrane fuel cell power module. International Journal of Hydrogen Energy 2016;41(31):13831–42.

[60] Burheim Odne S, Crymble Gregory A, Bock Robert, Hussain Nabeel, Pasupathi Sivakumar, du Plessis Anton, le Roux Stephan, Seland Frode, Su Huaneng, Pollet Bruno G. Thermal conductivity in the three layered regions of micro porous layer coated porous transport layers for the pem fuel cell. International Journal of Hydrogen Energy 2015;40(46):16775–85.

[61] Sandrock Gary. State-of-the-art review of hydrogen storage in reversible metal hydrides for military fuel cell applications. DTIC document, 1997.

[62] Lototskyy M, Satya Sekhar B, Muthukumar P, Linkov V, Pollet BG. Niche applications of metal hydrides and related thermal management issues. Journal of Alloys and Compounds 2015;645:S117–22.

[63] Czaja Alexander U, Trukhan Natalia, Müller Ulrich. Industrial applications of metal–organic frameworks. Chemical Society Reviews 2009;38(5):1284–93.

[64] Schlemminger C, Næss E, Bünger U. Cryogenic adsorption hydrogen storage with enhanced heat distribution – an in-depth investigation. International Journal of Hydrogen Energy 2016;41(21):8900–16.

[65] Schlemminger C, Naess E, Bünger U. Adsorption hydrogen storage at cryogenic temperature – material properties and hydrogen ortho-para conversion matters. International Journal of Hydrogen Energy 2015;40(20):6606–25.

[66] Carrette L, Friedrich KA, Stimming U. Fuel cells–fundamentals and applications. Fuel Cells 2001;1(1):5–39.

[67] Grove William Robert. XXIV. On voltaic series and the combination of gases by platinum. The London and Edinburgh Philosophical Magazine and Journal of Science 1839;14(86):127–30.

[68] Schönbein Christian Friedrich. X. On the voltaic polarization of certain solid and fluid, substances: to the editors. Philosophical Magazine and Journal 1839. http://dx.doi.org/10.1080/14786443908649684.

[69] Davy. An account of a method of constructing simple and compound galvanic combinations, without the use of metallic substances, by means of charcoal and different fluids. The Philosophical Magazine: Comprehending the Various Branches of Science, the Liberal and Fine Arts, Agriculture, Manufactures, and Commerce 1802;11(44):340–1.

[70] El-Kharouf A, Chandan A, Hattenberger M, Pollet BG. Proton exchange membrane fuel cell degradation and testing: review. Journal of the Energy Institute 2012;85(4):188–200.

[71] Burheim O, Vie PJS, Pharoah JG, Kjelstrup S. Ex-situ measurements of through-plane thermal conductivities in a polymer electrolyte fuel cell. Journal of Power Sources 2010;195:249–56.

[72] Bvumbe Tatenda J, Bujlo Piotr, Tolj Ivan, Mouton Kobus, Swart Gerhard, Pasupathi Sivakumar, Pollet Bruno G. Review on management, mechanisms and modelling of thermal processes in pemfc. Hydrogen and Fuel Cells 2016;1(1):1–20.

[73] Sharma Surbhi, Pollet Bruno G. Support materials for pemfc and dmfc electrocatalysts – a review. Journal of Power Sources 2012;208:96–119.

[74] Burheim O, Lampert H, Pharoah JG, Vie PJS, Kjelstrup S. Through-plane thermal conductivity of PEMFC porous transport layers. Journal of Fuel Cell Science and Technology 2011;8:021013.

[75] Bujlo P, Pasupathi S, Ulleberg Ø, Scholta J, Nomnqa MV, Rabiu A, Pollet BG. Validation of an externally oil-cooled 1 kw el ht-pemfc stack operating at various experimental conditions. International Journal of Hydrogen Energy 2013;38(23):9847–55.

[76] Ellamla Harikishan R, Staffell Iain, Bujlo Piotr, Pollet Bruno G, Pasupathi Sivakumar. Current status of fuel cell based combined heat and power systems for residential sector. Journal of Power Sources 2015;293:312–28.

[77] Burheim Odne, Vie Preben JS, Møller-Holst Steffen, Pharoah Jon, Kjelstrup Signe. A calorimetric analysis of a polymer electrolyte fuel cell and the production of H_2O_2 at the cathode. Electrochimica Acta 2010;55(3):935–42.

[78] Ochal Piotr, de la Fuente Jose Luis Gomez, Tsypkin Mikhail, Seland Frode, Sunde Svein, Muthuswamy Navaneethan, Rønning Magnus, Chen De, Garcia Sergio, Alayoglu Selim, et al. Co stripping as an electrochemical tool for characterization of Ru@Pt core-shell catalysts. Journal of Electroanalytical Chemistry 2011;655(2):140–6.

[79] Brandon Nigel P, Skinner S, Steele Brian CH. Recent advances in materials for fuel cells. Annual Review of Materials Research 2003;33(1):183–213.

[80] Larminie James, Dicks Andrew, McDonald Maurice S. Fuel cell systems explained, vol. 2. UK: J. Wiley Chichester; 2003.

[81] Chandan Amrit, Hattenberger Mariska, El-Kharouf Ahmad, Du Shangfeng, Dhir Aman, Self Valerie, Pollet Bruno G, Ingram Andrew, Bujalski Waldemar. High temperature (ht) polymer electrolyte membrane fuel cells (pemfc) – a review. Journal of Power Sources 2013;231:264–78.

[82] Sammes Nigel, Bove Roberto, Stahl Knut. Phosphoric acid fuel cells: fundamentals and applications. Current Opinion in Solid State & Materials Science 2004;8(5):372–8.

[83] Seland F, Berning T, Børresen B, Tunold R. Improving the performance of high-temperature pem fuel cells based on pbi electrolyte. Journal of Power Sources 2006;160(1):27–36.

[84] Hu Lan, Rexed Ivan, Lindbergh Göran, Lagergren Carina. Electrochemical performance of reversible molten carbonate fuel cells. International Journal of Hydrogen Energy 2014;39(23):12323–9.

[85] Alberti G, Casciola M, Massinelli L, Bauer B. Polymeric proton conducting membranes for medium temperature fuel cells (110–160°C). Journal of Membrane Science 2001;185(1):73–81.

[86] Bauer F, Denneler S, Willert-Porada M. Influence of temperature and humidity on the mechanical properties of Nafion 117 polymer electrolyte membrane. Journal of Polymer Science Part B: Polymer Physics 2005;43(7):786–95.

[87] Porada S, Zhao R, van der Wal A, Presser V, Biesheuvel PM. Review on the science and technology of water desalination by capacitive deionization. Progress in Materials Science 2013;58(8):1388–442.

[88] Wang Xingpu, Xu Ruoyu, Wang Rongfang, Wang Hui, Brett Dan JL, Pollet Bruno G, Ji Shan. Nano-sized co/co (oh) 2 core-shell structure synthesized in molten salt as electrode materials for supercapacitors. Ionics 2016:1–6.

[89] Burheim Odne S, Aslan Mesut, Atchison Jennifer S, Presser Volker. Thermal conductivity and temperature profiles in carbon electrodes for supercapacitors. Journal of Power Sources 2014;246:160–6.

[90] Hunter RJ. Foundation of colloid science. Oxford: Clarendon Press; 1993.

[91] Brogioli D. Extracting renewable energy from a salinity difference using a capacitor. Physical Review Letters 2009;103:058501.

[92] Sales BB, Saakes M, Post JW, Buisman CJN, Biesheuvel PM, Hamelers HVM. Direct power production from a water salinity difference in a membrane-modified superca-pacitor flow cell. Environmental Science & Technology 2010;44:5661–5.

[93] Biesheuvel PMD, Bazant MZ. Nonlinear dynamics of capacitive charging and desalination by porous electrodes. Physical Review E 2010;81:031502.

[94] Hauge HH, Presser V, Burheim O. In-situ and ex-situ measurements of thermal conductivity of supercapacitors. Energy 2014;78:373–83.

[95] Ricketts BW, Ton-That C. Self-discharge of carbon-based supercapacitors with organic electrolytes. Journal of Power Sources 2000;89(1):64–9.

[96] Ma Sang-Bok, Nam Kyung-Wan, Yoon Won-Sub, Yang Xiao-Qing, Ahn Kyun-Young, Oh Ki-Hwan, Kim Kwang-Bum. A novel concept of hybrid capacitor based on manganese oxide materials. Electrochemistry Communications 2007;9(12):2807–11.

[97] Brousse Thierry, Bélanger Daniel. A hybrid fe3 o 4 mno2 capacitor in mild aqueous electrolyte. Electrochemical and Solid-State Letters 2003;6(11):A244–8.

[98] Richter Frank, Kjelstrup Signe, Vie Preben JS, Burheim Odne S. Thermal conductivity and internal temperature profiles of Li-ion secondary batteries. Journal of Power Sources 2017;359:592–600.

[99] Burheim Odne S, Su Huaneng, Hauge Hans Henrik, Pasupathi Sivakumar, Pollet Bruno G. Study of thermal conductivity of pem fuel cell catalyst layers. International Journal of Hydrogen Energy 2014;39(17):9397–408.

INDEX

Printed in the United States
By Bookmasters